ANIMAL SCIENCE, ISSUES AND PROFESSIONS

HORSES

BIOLOGY, DOMESTICATION, AND HUMAN INTERACTIONS

Animal Science, Issues and Professions

Additional books in this series can be found on Nova's website under the Series tab.

Additional E-books in this series can be found on Nova's website under the E-books tab.

ANIMAL SCIENCE, ISSUES AND PROFESSIONS

HORSES

BIOLOGY, DOMESTICATION, AND HUMAN INTERACTIONS

JAMES E. LEFFHALM
EDITOR

Nova Science Publishers, Inc.
New York

For permission to use material from this book please contact us:
Telephone 631-231-7269; Fax 631-231-8175
Web Site: http://www.novapublishers.com

Additional color graphics may be available in the e-book version of this book.

LIBRARY OF CONGRESS CATALOGING-IN-PUBLICATION DATA
Horses : biology, domestication, and human interactions / editor: James E. Leffhalm.
p. cm.
Includes index.
ISBN 978-1-61761-303-6 (hardcover)
1. Horses. 2. Horses--Breeding. I. Leffhalm, James E.
SF285.H819 2010
636.1--dc22
2010026950

Published by Nova Science Publishers, Inc. † New York

Contents

PREFACE

This book presents topical research in the study of horses, including recent advances in equine herpes virus 4 genetics using BAC technology; selective breeding in horses; the sexual interference behavior in a managed herd of Sorraia horses; the management of wild horses with porcine zona pellucida; pathogen inactivation during horse manure composting; horse and human personalities; as well as horse identification and parentage control.

Chapter 1 - The equine herpesviruses are major infectious pathogens that threaten equine health. Equine herpesvirus 4 (EHV-4) is an important equine pathogen that causes respiratory tract disease, known as rhinopneumonitis, among horses worldwide. It causes significant economic losses to the equine industry because of respiratory problems and lost time for training and performing. EHV-4 genome manipulation with subsequent understanding of the viral gene functions has always been difficult due to the limited number of susceptible cell lines and the absence of small-animal models of the infection. Efficient generation of mutants of EHV-4 would significantly contribute to the rapid and accurate characterization of the viral genes. This problem has been solved recently by the cloning of the genome of EHV-4 as a stable and infectious bacterial artificial chromosome (BAC) without any deletions of the viral genes. Very low copy BAC vectors are the mainstay of present genomic research because of the high stability of inserted clones and the possibility of mutating any gene target in a relatively short time. Manipulation of EHV-4 genome is now feasible using the power of BAC technology, and should aid greatly in assessing the role of viral genes in the virus-host interaction.

Chapter 2 - In ancient world the horses were kept because they provided food as well as materials for clothing and shelter for our ancestors. In the modern world the most part of the horses are kept as a hobby and they are participated in racing, jumping, showing and etc. The horse's physical structure is the result of evolution, controlled initially by environment as well as selective breeding enforced by human. Evidence of selective breeding of horses for both riding and draught purposes exists from ancient Elam in 2800 BC. Development of the racing horses has been founded based on the fact that a horse's ability to run a given distance in a relatively short time is partly determined by genetic components. In selective breeding we need to define a breeding goal and some criteria to measure the performance of an individual animal. In race horses the criteria used to estimate the racing performance are time traits, handicap weight, performance rate, rank at finish, earning prize and the like. In this chapter the authors want to describe the history of the selection in horses, briefly, and to identify the criteria which have been used for measuring of the performance of the racing horses and might be used to maximize improvement of racing performance as well as to introduce some other criteria that can be used in selecting of the parents of racing horses, also, investigate the advantages and disadvantages of these criteria.

Chapter 3 - The functions of sexual interference behaviour were studied in a managed herd of Sorraia horses. The authors assessed the influence of age, kinship, affiliative relationships, dominance relationships, aggressiveness, oestrus state, and reproductive state on interference patterns to test five functional hypotheses: Sexual competition, mare protection, protection of a bond, dominance, and revanchism. Sexual interference behaviours consisted mainly of agonistic interactions directed by mares to the stallion, causing interruption of the sexual behaviour. On average, 29.3% of mating attempts suffered interference. Current reproductive state seemed to be the most important factor affecting rates of interference performed by mares: Barren mares interfered at higher rates than mares that foaled. Sexual interference decreased success of mating attempts, but neither decreased foaling success of receivers nor increased foaling success of interferers in the following year. The mare protection hypothesis received the strongest support from the data, as we found evidence that mares perceived mating as a threat and interfered to protect genetically related females that were involved. However, no single hypothesis tested was completely supported by the data. Future studies should test alternative hypotheses for the function of sexual interference behaviour as well as assess the effect of reproductive management practices on interference rates.

Chapter 4 - The advent of immunocontraception with porcine zona pellucida (PZP) has all but revolutionized wild horse management, providing a more humane method of population control than earlier strategies. Early studies on Assateague Island National Seashore have described it as an ideal form of fertility control in that it reduces the chance of conception to below 10%, can be delivered remotely, is reversible (after short-term use), lacks debilitating physiological side effects, cannot pass through the food chain, and shows minimal effects on social behaviors. However, recent research in other populations has revealed behavioral and physiological side effects of long-term PZP use. These results indicate that studies from one population may not necessarily be applicable to another, regardless of similarities in habitat and population structure. Careful study of the animals' demography, physiology, and behavior is necessary prior to and during treatment to ensure that a) the potential effects of PZP can be assessed accurately, and b) within managerial constraints, PZP effects are ameliorated as much as is possible. Here I explain the history of PZP use in wild horses, its side effects on the horses of Shackleford Banks, North Carolina in comparison to other populations, and offer management suggestions based upon wild horse biology and behavioral ecology, which may minimize or prevent these side effects in other populations.

Chapter 5 - Over the past few decades, composting has become one of the most common technological alternatives for horse manure valorization. However, such practices can lead to increased contamination of the farm environment to a greater extent and constitute a risk of infection for both humans and animals, especially if compost is spread in the fields where horses graze. After reviewing bacterial, viral and parasitic pathogens that can be found in horse manure, the authors shall describe processes that inactivate pathogens during composting, time-temperature regulations and guidelines applicable to horse manure composting operations. They shall then describe factors that could influence the composting processes and different ways that pathogens can be spread in the environment, before evaluating the risk that they represent for horse and human health. Finally, they shall discuss the relevance of international compost quality standards from a practical point of view.

Chapter 6 - Indigenous microbiota is a complex community of microorganisms that colonizes the whole gastrointestinal tract, some parts of other tracts and the air-exposed surfaces of animals. Generally, this community confers protection against the onset of pathogen infection and also provides many others benefits such as immune modulation or production of vitamins. Gastrointestinal microbiota is by far the best characterized microbiota in different animal species and in the human being but there is an increasing interest in other groups of microorganisms, such as those associated to the genitourinary tract.

Composition of these microbial communities varies between species, individuals and even within the different parts of a single organ. For example, the microorganisms of different portions of the enteric tract are strikingly different. Now it is also known that there is a host and site specific microbiota composition, also related to other factors such as age or health condition.

Probiotics are live microorganisms that when administered in adequate amounts confer a health benefit to the host. Preferably they should be native isolates, so the research in native microbiota acquires a significant relevance since they can be sources of potential beneficial microorganisms.

So far, equine native microbiota has not been extensively characterized. These microbial communities are related to the normal functions of different organisms and its characterization may provide important data about beneficial microbes useful for the promotion of horses' health.

In equines, probiotics could be used with different purposes, including improvement of the diet digestibility and prevention of metabolic or infectious disorders. In mares, probiotics could improve milk quality and quantity and microorganisms isolated from the vaginal tract could be directed to the prevention of fertility problems related to endometriris, a serious reproductive problem in the horse. Probiotics for foals would be useful for growth promotion and control of the problems associated with diarrheas. Also, probiotics-based strategies could be designed for specific groups of animals, such as racing horses.

In this chapter, equine gastrointestinal and vaginal microbiota are reviewed and attention is focused on modulation strategies based on the use of probiotics for health promotion.

Chapter 7 - Scientific publications on animal behavior give increasing attention to personality, regarded in terms of consistent behavioral differences between individuals. And variability in behavioral traits is hypothesized to have important ecological and evolutionary implications. Working from considerations of the origins of domestication, the authors inquire into the subtle though lasting effect of interacting personalities from both sides of the horse-human relationship.

The early human's capability for horse-keeping developed through a gradual build-up of knowledge on the horse's social organization and basic needs. So at the early stages of horse domestication, humans would have needed to overcome certain cognitive and logistical difficulties as a prerequisite to controlled horse breeding. And this would have required people holding particular skills and possessing particular personality traits.

Since personality traits are the result of natural processes shaped by genetic and ontogenetic determinants, they are eligible as significant factors on both sides of the horse-human relationship. At the horse side, we can suppose that particularly amenable individuals were selectively and repetitively bred at key breeding locations in the emergence of horse lineages. At the human side, the authors suggest that conditions would have been conducive to development of an in-depth understanding of the horse's behavior and abilities, and that this knowledge would have been harnessed to devise training schemes to make the most of the horse's true capabilities.

Chapter 8 - Accurate determination of relatedness is of great importance in the horse breeding industry. In the past decade, the DNA forensic field was primarily focused on human genome typing, for identification and parentage determination purposes. The discovery of polymorphisms in microsatellites loci (STRs) and the introduction of polymerase chain reaction technique has lead to the establishment of an extremely powerful method for identification and for parentage analysis also in animals. Microsatellites have been in fact identified in all eukaryotic species studied so far and modern tools of molecular biology provided means for a fast and accurate way for animal genotyping. Currently, the majority of horse breed registries adopted DNA technology in order to verify pedigree records and to solve queries of parentage.

Chapter 9 - Horse identification and parentage control are established in all countries where equestrian sports are developed. As in human forensics, the horse identity can be confirmed by DNA typing.

Blood, saliva, hair roots and tissue fragments provide a reliable source of DNA, but, when it is impossible to recover those biological samples, DNA can be extract from "challenging" samples, such as bone, tooth and urine samples.

Chapter 10 - The Wild Free-Roaming Horses and Burros Act of 1971 (Public Law 92-195) provided protection for all wild horses and burros on federal lands and provided guidance for their management as a wildland species. At the time the Act was passed, a roughly estimated 17,000 wild horses1 occupied federal lands designated for the protection. Since 1971, the primary responsibility for management of the wild equids on federal lands has fallen primarily to the Bureau of Land Management (BLM), with the U.S. Forest Service sharing some responsibility. In 1971, the BLM was neither staffed nor prepared to monitor and manage such a large number of horses and their effects on semiarid ecosystems. By 1980, the number of wild horses had increased to 65,000-80,000 animals.

[1] Since no systematic surveys were conducted at this time, some authors feel this number may not be representative of the numbers in 1971.

In: Horses: Biology, Domestication, and Human Interactions
Editor: James E. Leffhalm
ISBN: 978-1-61761-303-6

Chapter 1

Recent Advances in Equine Herpesvirus 4 Genetics Using BAC Technology

Walid Azab and Hiroomi Akashi
Department of Veterinary Microbiology, Graduate School of Agricultural and Life Sciences, The University of Tokyo, Bunkyo-ku, Tokyo, Japan

Abstract

The equine herpesviruses are major infectious pathogens that threaten equine health. Equine herpesvirus 4 (EHV-4) is an important equine pathogen that causes respiratory tract disease, known as rhinopneumonitis, among horses worldwide. It causes significant economic losses to the equine industry because of respiratory problems and lost time for training and performing. EHV-4 genome manipulation with subsequent understanding of the viral gene functions has always been difficult due to the limited number of susceptible cell lines and the absence of small-animal models of the infection. Efficient generation of mutants of EHV-4 would significantly contribute to the rapid and accurate characterization of the viral genes. This problem has been solved recently by the cloning of the genome of EHV-4 as a stable and infectious bacterial artificial chromosome (BAC) without any deletions of the viral genes. Very low copy BAC vectors are the mainstay of present genomic research because of the high stability of inserted clones and the possibility of mutating any gene target in a relatively short time. Manipulation of EHV-4 genome is now feasible using the power of BAC technology, and should aid greatly in assessing the role of viral genes in the virus-host interaction.

I. Introduction

In equidae, nine herpesviruses have been isolated. Six belong to the subfamily alphaherpesvirinae and three to the subfamily gammaherpesvirinae. The equine herpesviruses type 1 (EHV-1), 3 (EHV-3), 4 (EHV-4), 6 (EHV-6), 8 (EHV-8), and 9 (EHV-9) are typical alphaherpesviruses whereas the equine herpesviruses type 2 (EHV-2), 5 (EHV-5), and 7 (EHV-7) are gammaherpesviruses [48]. Horses are the natural host to EHV-1, EHV-2, EHV-

3, EHV-4, and EHV-5 while donkeys are the host to EHV-6 (asinine herpesvirus 1, AHV-1), EHV-7 (AHV-2), and EHV-8 (AHV-3). Thomson's gazelles are the host to the newest member of the equine herpesviruses EHV-9 (gazelle herpesvirus) [48, 69].

EHV-4 is a major cause of respiratory disease in horses throughout the world. It causes significant economic losses to the equine industry because of respiratory problems and lost time for training and performing. Infections of horses with EHV-4 are clinically difficult to be distinguished from those caused by EHV-1, the close relative of EHV-4. Although both viruses cause respiratory disease, only EHV-1 has been established as a major cause of epidemic and sporadic abortion, prenatal mortality and neurological disorders that can range in severity but often including complete paralysis [4, 144, 170].

EHV-4 genome manipulation with subsequent understanding of the viral gene functions has always been difficult due to the limited number of susceptible cell lines and the absence of small-animal models of the infection. Efficient generation of mutants of EHV-4 would significantly contribute to the rapid and accurate characterization of the viral genes. More recently, many laboratories reported the cloning of large herpesvirus genomes as bacterial artificial chromosomes (BACs) [1, 9, 10, 12, 21, 32, 52, 53, 85, 115, 120, 134, 154, 161, 173, 186, 192, 211, 215]. This technique allows the maintenance and mutagenesis of herpesvirus genomes in *Escherichia coli* (*E. coli*). The previous requirement for selection and rounds of plaque purification in eukaryotic cells to generate recombinants are no longer required [74]. Targeted and random mutagenesis can be introduced easily into the viral genomes while being maintained in *E. coli*. The reconstitution of progeny virions is achieved by transfection of the BAC plasmid into permissive mammalian cells, thereby allowing the generation of mutant viruses more quickly [115].

II. General Characteristics of Equine Herpesvirus 4

A. Classification

The new order *Herpesvirales* is divided into three families; the revised family *Herpesviridae* retains the mammal, bird and reptile viruses, the new family *Alloherpesviridae* incorporates the fish and frog viruses, and the new family *Malacoherpesviridae* contains a bivalve virus [48]. EHV-4 and its close relative EHV-1 are members of the *Alphaherpesvirinae* subfamily, genus *Varicellovirus* [48, 151]. The complete genome sequences of EHV-4 and EHV-1 were reported [188, 189]. Both viruses have a linear double stranded type D DNA genome of approximately 145 (EHV-4) and 150 (EHV-1) kilobasepair (kbp) and a similar (57 %) G + C content [45, 46, 152]. The genomes are composed of a unique long (U_L) and a unique short (U_S) sequences flanked by inverted repeat regions (TR_S/IR_S) [189]. Both genomes have 76 unique corresponding genes with a potential to code for 77 different proteins due to spliced expression of two genes. However, the actual number of open reading frames (ORFs) in the genomes is 79 and 80, respectively, for EHV-4 and EHV-1 due to duplication of three (EHV-4) and four (EHV-1) genes. Sequence analysis showed that there is from 55–84 % DNA homology depending on the gene and that the homology at the amino acid level ranged from 55–96 % [189].

B. Epidemiology and Pathogenesis

Both EHV-4 and EHV-1 are endemic in horse populations worldwide and the majority of horses show serologic evidence of exposure to theses viruses [4, 144]. However, the extensive antigen cross-reactivity between the two viruses does not allow differentiation between them. The development of type-specific serologic tests has in recent years allowed investigation of the seroprevalence of EHV-1 and EHV-4 infections in different populations of horses and in different age categories [43].

Primary infection with either EHV-1 or EHV-4 occurs via direct horse-to-horse contact and inhalation of aerosolized virus. Direct contact with EHV-1 infected aborted fetuses, placental tissues or fomites contaminated by respiratory secretions may serve as a source of infection; however, the virus is short-lived in the environment [75, 149]. It is assumed that reactivation of EHV-4 from latency causes recurrent disease and is accompanied by virus shedding and transmission either directly or indirectly to other horses in the herd. In horses, both viruses spread by the respiratory route, with primary replication occurring in respiratory epithelia, resulting in acute respiratory manifestations. However, EHV-1 is associated with epidemic and sporadic abortion, prenatal mortality and neurological disorders [4, 43, 56, 57].

The pathogenicity of EHV-1 is ascribed to the capacity of the virus to rapidly reach the lymphatic tissues associated with the upper respiratory tract, and to infect mononuclear cells that ultimately reach the bloodstream and lead to cell-associated viremia [107, 205]. Carried by the infected peripheral blood mononuclear cells (PBMC), EHV-1 can reach the vasculature of other tissues such as the pregnant uterus or the central nervous system, where the virus can attach to, enter and replicate in endothelial cells. Although respiratory infections of horses with EHV-4 are clinically difficult to be differentiated from those caused by EHV-1, infection with EHV-4 has been rarely associated with abortions or paresis [116, 144, 170]. The differences in pathogenicity between EHV-1 and EHV-4 seem to be reflected in the host range for both viruses in various cultured cells. EHV-1 can propagate well in many cell lines including, equine, bovine, rabbit, hamster, mouse, monkey, pig and cat derived cell lines [203] while EHV-4 appears to be restricted mainly to equine-derived cells [114].

C. Clinical Disease and Diagnosis

EHV-4 is a major cause of acute upper respiratory disease, with most horses being infected during the first two years of life. EHV-4 is one of at least nine different viruses known or suspected to cause respiratory disease during this time of life of the horse. It is noteworthy that there are no distinguishing characteristics between EHV-1 and EHV-4 respiratory disease [43, 75]. The incubation period for the disease is between 3 and 10 days. This is slightly longer than that reported for equine influenza virus, 1 to 5 days, which may help distinguish between the two diseases. Acute naturally occurring EHV-4 respiratory disease is characterized by fever, depression, lethargy, anorexia, ocular discharge, and profuse serous nasal discharge that later becomes mucopurulent [184]. Lesions occur on the mucus membranes of the upper, sometimes lower, respiratory tract and may also involve the lungs. In uncomplicated cases, clinical signs persist for 2 to 7 days. However, both upper and lower respiratory forms are frequently complicated by secondary bacterial infections which considered a contributory factor in the development of rhinopneumonitis [43].

Preliminary diagnosis can be done based on the presence of compatible clinical signs and the rapid spread of the disease. Early hematological analyses include mild anemia and lymphopenia, which are followed by mild hyperfibrinogenemia. If affected horses develop secondary bacterial infection, leukocytosis with a mature neutrophilia, marked hyperfibrinogenemia, and hyperglobulinemia were observed [75]. Traditionally, virus isolation, identification and typing are the gold standard methods for diagnosis. The sample of choice is a nasopharyngeal swab which should be taken early in the febrile stage of the disease. For virus isolations, the samples are prepared and inoculated into monolayer cultures of equine cells [197, 203].

PCR offers an alternative to virus isolation and has proven to be a sensitive and rapid method for identification of EHV-4 DNA in respiratory secretions [163]. PCR has also detected viruses that was no longer cultivable in cell culture and has been used to amplify DNA from formalin-fixed tissues or from stained tissue sections from histological slides [139]. Alternative molecular approaches have been established using real-time TaqMan PCR which reduce the risk of contamination and allow discrimination between the different viral states [145]. PCR results are generally obtained within 24 to 48 hours of sample processing, allowing early implementation of infectious disease control measures to reduce the spread of the virus.

Due to the high seroprevalence of EHV-4 and the antigenic cross-reactivity between EHV-1 and EHV-4, specific serodiagnosis is only possible retrospectively by showing significant increase in titer between acute and convalescent serum samples [4, 43]. Type-specific serologic assays have been developed that allow the clear distinction of EHV-4, EHV-1, and dual-infected horses [41, 42]. These assays detect antibodies to a strong, type-specific epitope (s) on the glycoprotein G of EHV-4 and EHV-1. Although these assays are not routinely offered by commercial laboratories, many veterinary laboratories now offer routine serodiagnostics for EHV-4 and EHV-1.

D. Latency

Latency by alphaherpesviruses is an important epidemiological strategy ensuring survival and spread within the natural host population [150]. For both EHV-1 and EHV-4 latency has been identified in neuronal cells of the trigeminal ganglia as well as in lymphoid tissues. While the $CD5^+/CD8^+$ T lymphocytes were defined as the predominant site of EHV-1 latency, EHV-4 has only rarely been isolated from peripheral blood lymphocyte [19, 20, 58, 172].

The physiological factors responsible for the reactivation of latent viruses remain undefined. However, it is well recognized that spontaneous shedding may follow weaning, castration, relocation, terminal illness and treatment with corticosteroids and cyclophosphamides [25, 31].

Table 1. Glycoproteins of EHV-1 and EHV-4

EHV-1 and EHV-4		Homologue	Functions
Gene	Name	HSV-1	
6	gK	UL53	Nonessential, virus egress and cell-to-cell spread*
16	gC	UL44	Nonessential, role in cell entry
33	gB	UL27	Essential, role in cell entry
39	gH	UL22	Essential, complexes with gL; role in cell entry
52	gM	UL10	Nonessential, virus egress and cell-to-cell spread
62	gL	UL1	Essential, complexes with gL; role in cell entry
70	gG	US4	Nonessential, chemokine binding activity*
71	gp2	US5†	Nonessential, virus egress and cell-to-cell spread*
72	gD	US6	Essential, role in cell entry
73	gI	US7	Nonessential, cell-to-cell spread
74	gE	US8	Nonessential, cell-to-cell spread

*Functions were assigned for EHV-1 only as there is no previous data with EHV-4 [87, 138, 155, 193].
†Positional counterpart only [189].

D. Host Cell Tropism

The glycoproteins of herpesviruses play important roles in the pathogenicity, the infection process (virus adsorption, entry, penetration and cell-to-cell spread) and as targets by virtue of their expression on the surface of infected cells for the natural host's immune responses [178]. Accordingly, the glycoproteins and their genes have been the best studied of the EHV-4 and EHV-1 proteins. Homologues for at least 10 [glycoprotein B (gB), gC, gD, gE, gG, gH, gI, gK, gL, and gM] of the 11 HSV-1 glycoproteins of both EHV-4 and EHV-1 have been identified [189] (Table 1). Both EHV-4 and EHV-1 have additional glycoproteins which have no structural or functional counterparts in HSV-1 [43].

Some alphaherpesviruses, such as HSV-1 and pseudorabies virus (PrV), infect a broad range of cultured cells [179]. This appears to be the case with EHV-1, but not EHV-4, which can be easily propagated in many cell lines derived from different species [4, 43, 203].

As the horse is an inconvenient animal for experimental studies of EHV-4 and EHV-1, the availability of laboratory animal models is particularly important. Although animal models were found to be non-permissive for EHV-4 replication, EHV-1 can establish replication and produce infection in hamsters and mice [15, 43, 183].

Difference in cell tropism may well be associated with differences in the gene products involved in virus entry and/or spread from an infected to a neighboring uninfected cell. Such functions are regularly executed by herpesviral envelope glycoproteins. Studies with HSV-1, the prototypical alphaherpesvirus, have shown that basically five viral glycoproteins (gC, gB, gD, gH, and gL) and a cellular receptor (HveA, nectin-1, or 3-O- sulfated heparan sulfate) are required to efficiently initiate productive infection. Virions first attach to target cells using gC to bind the heparan sulfate moieties of the cell surface proteoglycans [78, 165]. However, gC is not essential for viral replication and in its absence gB can mediate the attachment step [80]. The envelope gD is specific to alphaherpesviruses and is essential for entry of most of them [94, 109]. gD is necessary for stable binding of the virus to cell surface receptors, thereby triggering fusion. The most widely used gD receptors are nectin-1, nectin-2, HveA, or

3-O- sulfated heparan sulfate [129, 168, 181]. Finally, gB in combination with gD and the gH/gL complex is responsible for the fusion of the viral envelope with the plasma membrane and/or from within endosomes and the subsequent release of the virus core into the cytoplasm [180, 181]. Additional proteins have been reported to intervene during alphaherpesvirus entry. On the viral side, gK and gM modulate fusion process of HSV-1 and PrV [11, 47, 106, 124]. On the cellular side, a lipid-raft-associated receptor and paired immunoglobulin-like type 2 receptor (PILR) alpha have been suggested for HSV-1 gB [16, 17, 158], and cellular integrins have been shown to bind to HSV-1 gH/gL and EHV-1 gD [142, 194].

As is the case with other alphaherpesviruses, EHV-1 also enters cells through direct fusion at the plasma membrane at neutral pH, a process that is mediated by gB, gC, gD and presumably the gH/gL complex [44, 137, 141]. It was shown also by electron microscopy that EHV-1 can enter cells via a nonclassical, endocytic/phagocytic pathway [66]. Recently, it was described that EHV-1 enters PBMC predominantly via the endocytic pathway and that this entry is triggered by the interaction between cellular integrins and gD. Moreover, EHV-1 can use different cellular entry pathways to infect important target cell populations of its natural host [194]. Taken together, it has been suggested that the broad tissue tropism of EHV-1, but not EHV-4, may be due to its ability to use multiple cellular receptors and different pathways to initiate virus entry.

E. Immunity and Vaccination

The protective immune mechanisms against EHV-4 and EHV-1 infections remain poorly understood. This is particularly difficult because of the lack of immunologically and virologically naïve natural hosts. High levels of circulating virus neutralizing antibodies (VN) do not consistently correlate with protection from natural or experimental infection and viral shedding from the respiratory tract [77, 143]. As with other respiratory infections, mucosal antibodies and/or cell-mediated immunity may play a role in controlling EHV-1 and EHV-4 infections [24]. Cytotoxic T lymphocytes (CTLs) form an essential part of the immune defense against many virus infections. However, the role of CTLs in EHV-4 and EHV-1 infections is at the early stages.

Currently there are at least 12 multivalent EHV-1 and EHV-4 inactivated and modified live vaccines available in the market. However, these vaccines can not block the infection with equine herpesviruses and most of them only claim protection against respiratory diseases due to EHV-1 and EHV-4 [77, 144]. The most we can hope from using such vaccines is the reduction of severity of clinical signs, viremia, and attenuation of viral shedding to help protect the herd. Genetically engineered live virus vaccines, particularly those that have gene (s) either deleted and/or inserted, comprise the area of interest in herpesvirus vaccine development. Nonetheless, much effort has gone into developing live deletion mutant vaccines, mostly using EHV-1 strains. EHV-1 with at least six genes deletions [117] and a TK deficient strain of EHV-1 [38] have been constructed. Both strains produced poor or no VN antibody and the horses were not significantly protected when challenged four weeks later. Similar findings were also reported for a double (gE and gI) deletion mutant of EHV-1 [118]. Hence, unlike the case with BHV-1 [95] and PrV [127], no suitable candidate EHV-1 or EHV-4 deletion mutant to our knowledge has been generated.

F. Prevention and Control

For prevention and control of EHV-4 and EHV-1-related diseases, the following management practices are strictly recommended:

1. Isolation of the infected horses is the key procedure for controlling the disease. However, this should be done in well-ventilated stalls. Additionally, the in-contact horses should also be isolated and monitored for at least one month.
2. Vaccination of the unexposed horses can also reduce the risk of virus spread to other farms.
3. Horses arriving on a farm from other farms or sales or returning from the racetrack should be isolated for 3-4 weeks before commingling with resident horses.
4. Pregnant mares should be maintained in a group away from weanlings, yearlings, and horses out-of-training.
5. All organic materials should be removed from stalls, along with cleaning of contaminated surfaces with detergents.
6. Minimizing stress-associated reactivation and shedding of virus from latently infected carrier horses is highly recommended.

III. Cloning of Herpesviruses as BAC

Herpesviruses are important pathogens for animals and humans, and have the largest DNA genomes among mammalian viruses with up to 200 potential ORFs [151]. The high degree of adaptation to their hosts appears to be the cause of having these large and complex genomes. Herpesviruses can manipulate the virus-host interaction in different ways, such as choosing specific target cells, establishing latent infection, controlling cell-cycle and transforming cells. Functional characterization of these viral genes by generating virus mutants is important for understanding the molecular aspects of herpesvirus replication and pathogenesis, and to provide basis for the rationale development of new vaccines and chemotherapeutics. Although conventional methods for mutagenesis of herpesviruses, chemical mutagenesis, site-directed mutagenesis by homologous recombination in eukaryotic cells or cosmid technology, proved to be very useful for the generation of a variety of mutants, their construction is often inefficient, laborious and time consuming [27, 121, 201]. The limitations of previous methodologies for generation of herpesvirus mutants have stimulated the development of new cloning vehicles suitable for the maintenance of large DNA inserts. Yeast artificial chromosomes (YACs) were the first and undoubtedly the largest, with a coding capacity of more than 2 Mbp. However, YACs are not easy to handle and the recombination machinery of yeast renders YACs prone to unwanted rearrangements and deletions [147, 159].

Recently, *E. coli*-based cloning vectors depends on the replicons of bacterial F plasmids have been developed [125, 166] and, by analogy to YACs, named BACs. They can maintain foreign DNA sequences of up to 300 kbp and are easy to handle. The strict control of the F-factor replicons, by the regulatory functions of *E. coli*, maintains a single copy of the BAC per bacterial cells. This allows stable maintenance of large DNA inserts and reduces the risk of recombination events between DNA fragments carried by the vector [167]. These properties made BACs attractive for herpesvirologists, as even the largest known herpesvirus genome could be cloned as a single BAC. In 1997, Messerle and his colleague reported on the first infectious herpesvirus BAC, murine cytomegalovirus BAC [125]. Subsequently, the cloning of several herpesvirus genomes as BACs has been described [1, 9, 10, 12, 21, 32, 52, 53, 85, 115, 120, 134, 154, 161, 173, 186, 192, 211, 215].

BAC cloning is not simply a matter of ligating the viral genome into the BAC plasmid vector, since large herpesvirus genomes are not typically amenable to such manipulation. First, the BAC plasmid cassette is introduced into the herpesvirus genome by conventional homologous recombination in infected mammalian cells. The linear double-stranded DNA genome of herpesviruses circularizes during replication. It suffices to isolate the circular replication intermediate of the BAC mutant and to shuttle it by DNA transformation into competent *E. coli*. The herpesvirus BAC is then propagated and mutated in *E. coli*. Such a viral BAC would be infectious, as transfection of BAC-cloned viral DNA into permissive cells should lead to virus production [164] (Figure 1).

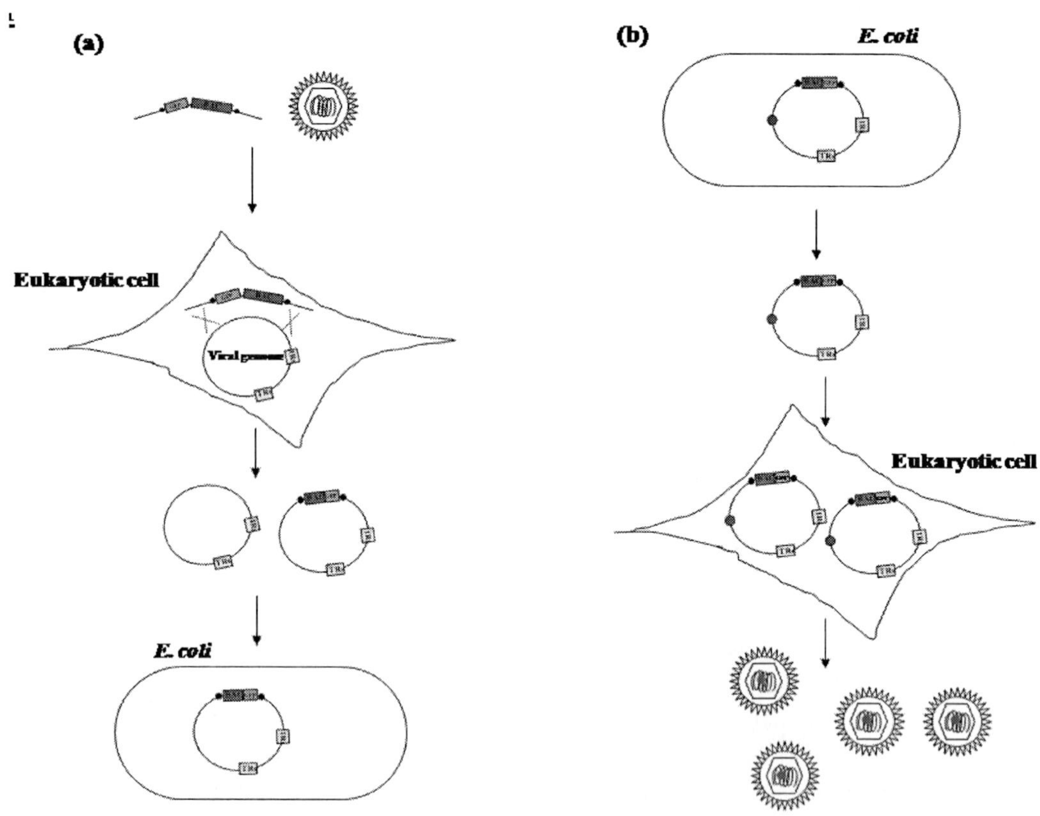

Figure 1. Cloning a viral genome as a bacterial artificial chromosome. (a) A mutant virus containing the BAC cassette is produced by homologous recombination in infected cells. Circular replication intermediates of the viral genome are isolated and transformed into *E. coli* to obtain a virus BAC. (b) The BAC-cloned viral genome can be engineered in *E. coli* to introduce a mutation (red). Transfection of a BAC containing the mutated viral genome leads directly to mutant viral progeny.

The application of BAC technology to herpesviruses has represented a significant advance in the study of the molecular genetics of these ubiquitous pathogens. BACs probably represent the most flexible and efficient technology to date for generating mutant viruses, and should aid greatly in defining the role of specific viral gene products in pathogenesis. The availability of cloned viral genomes as BACs could allow for rapid sequencing of the unsequenced herpesvirus genomes. The large size of herpesvirus genomes, with numerous nonessential genes, is one of the properties that make herpesviruses attractive for gene therapy. The generation of recombinant viruses, facilitated by BAC technology, expressing therapeutic genes instead of the nonessential genes should fuel interest in the use of these viruses as gene therapy vectors. Finally, BACs should prove useful in vaccine design, not only as a substrate upon which to generate safe live attenuated vaccines, but also as immunogens in their own right, as DNA vaccines [27, 185, 201].

A. Construction of an Infectious EHV-4 BAC Clone

In the case of EHV-4, the BAC cassette was inserted into the intergenic region between the 58 and 59 genes of the virus genome. This region was chosen for the integration of the BAC vector since it is predicted to be nonessential for viral replication [189]. Furthermore, the BAC vector sequences have been introduced into EHV-4 genome without any deletions or destruction of any of the viral genes. However, for many other herpesviruses, the BAC cassette has been inserted instead of one or more nonessential genes resulting in either noninfectious or infectious viruses with a deletion mutation [9, 32, 125, 154, 192]. Thus, to insert the BAC vector sequences into the genome of EHV-4, a transfer vector was constructed containing around 2.2 kbp fragments homologous to the flanking sides of the intergenic region of genes 58 and 59 (the desired integration site) and the BAC vector including GFP (Green Fluorescent Protein, served as selection and screening marker) and flanking *loxP* sites to allow the excision of the BAC cassette (Figure 2). The insertion of BAC plasmid cassette into EHV-4 genome was done typically by homologous recombination in primary fetal horse kidney (FHK) cells. This step was done by transfection of linearized transfer vector into FHK cells infected with EHV-4. The development of virus plaques showing green fluorescence indicated the integration and expression of the GFP gene. After 3 further rounds of plaque purification, GFP fluorescing recombinant virus, WA78, was obtained (Figure 3).

Generation of a stable BAC in *E. coli* next requires transformation of competent bacterial cells with circular replication intermediates of the BAC clone. Since the linear double stranded DNA genome of herpesviruses circularizes during replication, circular replication intermediate of the BAC clone can be isolated from infected cells soon after virus infection. Choosing the right time to extract circular viral DNA in sufficient quantity seems to be critical to obtaining a high proportion of full-length infectious clones [121, 201]. In the case of EHV-4, the circular intermediates were isolated from infected cells and electroporated into *E. coli* DH10B. Isolation of BAC DNAs from one chloramphenicol resistant colony and restriction enzyme analysis led to the identification of a bacterial clone containing a BAC plasmid with the full-length EHV-4 genome. Further analysis was done by PCR, Southern blot analysis, and sequencing of the targeted site.

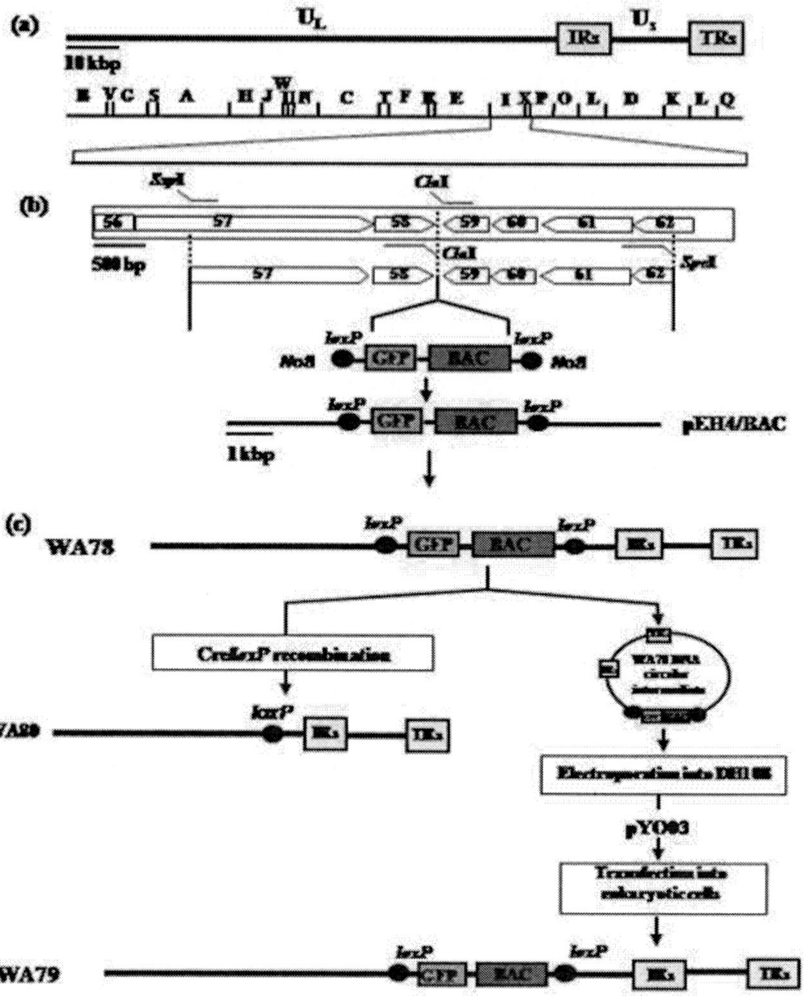

Figure 2. Schematic illustration of the strategy used to construct infectious EHV-4 BAC plasmid. (a) Schematic presentation of the genomic organization and the *Bam*HI restriction map of the entire EHV-4 TH20p strain. The two unique regions (U_L and U_S) and the terminal and internal repeat sequences (TR_S and IR_S) are shown. (b) Schematic presentation of the genomic domain that encodes the genes from 56 to 62 is depicted. The *loxP*-flanked BAC cassette was inserted into the intergenic region between genes 58 and 59 by homologous recombination in FHK cells. The transfer vector pEH4/BAC was used to generate WA78. The region of insertion is located in the *Bam*HI I restriction fragment. (c) Schematic presentation of the reconstitution of EHV-4 infectious clone and excision of BAC cassette. Circular intermediate form of WA78 was isolated from infected FHK cells and transformed into *E. coli* DH10B. The pYO03 BAC DNA was transfected into eukaryotic cells to reconstitute the virus WA79. The BAC cassette was excised by co-infection of FHK cells with WA78 and a recombinant adenovirus that expresses the Cre recombinase. As a result, the BAC cassette was excised and a single *loxP* site was remained in the intergenic region between genes 58 and 59 of the newly generated virus WA80.

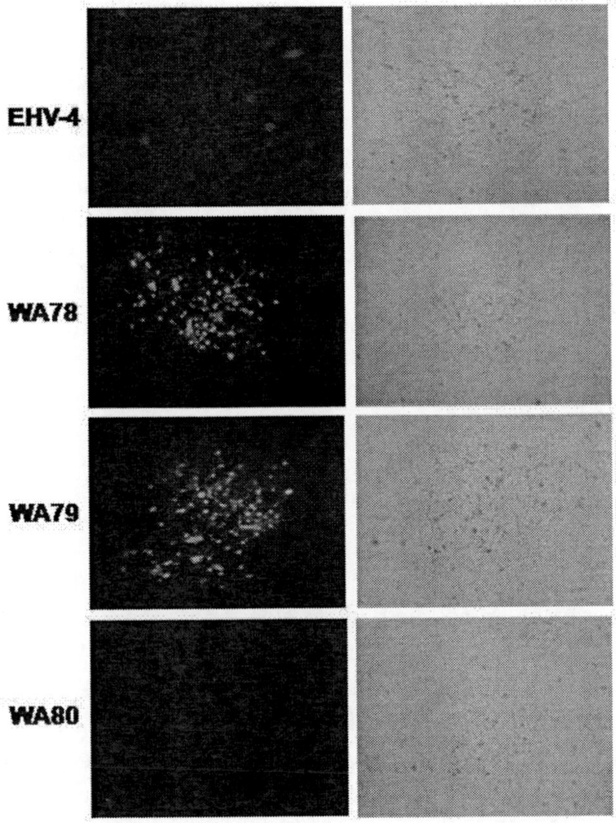

Figure 3. Plaque morphology of EHV-4 strains derived from EHV-4 BAC plasmid. FHK cells were infected with EHV-4, WA78, WA79, or WA80 and overlaid with DMEM containing 0.5 % methylcellulose. Three days later, the plaques were observed under a fluorescent microscope (left panel) or a light microscope (right panel).

B. Stability of the Viral Bac Clone in *E. Coli*

An important aspect of cloning viral genomes as BACs is the stability of the BAC clones in *E. coli* during maintenance, propagation and mutagenesis. Experimental studies with cloned mammalian and plant genomes revealed that BACs are remarkably stable when propagated in suitable bacterial strains [166]. Since herpesvirus genomes contain repetitive sequences that are targets for recombination, herpesvirus BACs are usually propagated in the *E. coli* strain DH10B, which carries a *recA* mutation that minimizes recombination [3, 201]. Hence the earlier experiences have shown that even mutations in nonessential genes often lead to virus attenuation, analyzing the biological fitness is probably a trustable indicator of the genomic stability. Stable propagation in *E. coli* DH10B has been shown for several herpesvirus BACs [21, 53, 84, 85, 115]. However, relative instability of such BAC plasmids has been reported for other herpesviruses, such as murine gammaherpesvirus 68 (MHV-68) which contains a large number of repeated sequences which prone to recombination events [1]. It was observed that there is a loss of some of the 40 bp repeat units of MHV-68.

However, this loss had no influence on the reconstitution of infectious virus and on the *in vitro* growth properties [1, 3].

For EHV-4 BAC genome, stability was demonstrated by serial propagation in the *E. coli* strain DH10B for 20 consecutive days. As *E. coli* has a generation time of around 20 min, and given that the stationary growth phase is reached at 12 h, each day represents approximately 36 generations [115]. Furthermore, bacteria with the BAC plasmid obtained after the last passage was plated on agar containing chloramphenicol and overnight cultures were grown from single colonies. After the various periods of culture, the BAC DNA was isolated and analyzed by restriction enzyme digestion. All BAC plasmids grown for various periods of time showed a restriction pattern that was similar to that of the original EHV-4 BAC (Figure 4), indicating a high stability of the EHV-4 BAC in *E. coli*.

C. Reconstitution of Infectious Viruses and Excision of the BAC Sequence from the Recombinant Viruses

In general, virus reconstitution is achieved by transfection of the viral BAC DNA into permissive cells either by chemical methods or electroporation. For EHV-4, the reconstitution of infectious virus, WA79, was performed by transfection of EHV-4 BAC DNA into human embryonic kidney cell line 293 (293 cells); these cells generally allow a high efficiency of transfection. GFP-expressing cells were observed as early as 16 h after transfection. After 2 days, the supernatants and cells were collected and used to infect confluent FHK cells. Viral plaques with green fluorescence were observed 2 or 3 days after infection as well as CPE was spread into the culture (Figure 3).

As mentioned above, cloning of herpesvirus genomes as BAC requires the insertion of the BAC vector sequences into the viral genome. However, the presence of the BAC vector sequences in the reconstituted viruses, especially when the BAC cassette was inserted instead of one or more nonessential genes, may interfere with the biological properties of these viruses either *in vitro* or *in vivo*. Several previous works had reported that the reconstituted viruses had different growth properties from the wild-type viruses *in vitro* [1, 9, 154]. However, there are few previous reports demonstrating that the reconstituted viruses from BAC clones have wild-type phenotypes in cell cultures without excision of the BAC cassette [74, 186, 213]. In the case of EHV-4, the reconstituted viruses from the BAC plasmid were indistinguishable from wild-type virus *in vitro* even without excision of the BAC sequence [12].

Observations that emphasize the need to remove the BAC sequences from the viral genome have been made with many BAC-cloned herpesviruses [1, 174, 186, 199, 211]. Thus, the design of the virus BAC-clone should be made in a way that allows easy and efficient excision of the BAC sequence. This could be achieved easily by exploiting the power of Cre recombinase-based strategy [135]. In this strategy, the BAC vector sequences, including GFP gene or any additional genes, must be flanked by two *loxP* sites. One of the great advantages of the Cre/*loxP* recombination system is that there is no need for additional co-factors or sequence elements for efficient recombination regardless of the cellular environment. Concerning the molecular mechanism of recombination, a single recombinase molecule binds to each palindromic half of a *loxP* site, then the recombinase molecules form a tetramer, thus bringing two *loxP* sites together [198]. The recombination occurs within the spacer area of the *loxP* sites. The postrecombination *loxP* sites are formed from the two complementary halves of the prerecombination sites. For some herpesviruses [174, 191, 211, 214], excision of the BAC DNA backbone was achieved by Cre recombinase which was encoded by the BAC vector itself. When these BAC plasmids were transfected into permissive eukaryotic cells, Cre was expressed and mediated site-specific recombination between the *loxP* sites, excising the BAC DNA sequences without the need for serial passage or plaque purification. In contrast to these viruses, where the BAC vector sequences are excised whenever the BAC-cloned genomes are delivered into eukaryotic cells, the BAC DNA backbone in the EHV-4 reconstituted viruses can be excised upon the need. This can be easily performed by infection of permissive mammalian cells with the recombinant adenovirus, AxCANCre, which expresses the Cre recombinase [96] followed by superinfection of these cells with EHV-4/BAC. Deletion of the BAC cassette from the obtained viruses, WA80, can be monitored by the absence of GFP expression (Figure 3). Like many other herpesviruses [2, 9, 154, 174, 186, 191, 211], a single 34 bp *loxP* site remains in EHV-4 viral genome after excision of the BAC vector sequences [12]. However, the virus growth in cell culture was not affected by the existence of these nucleotides.

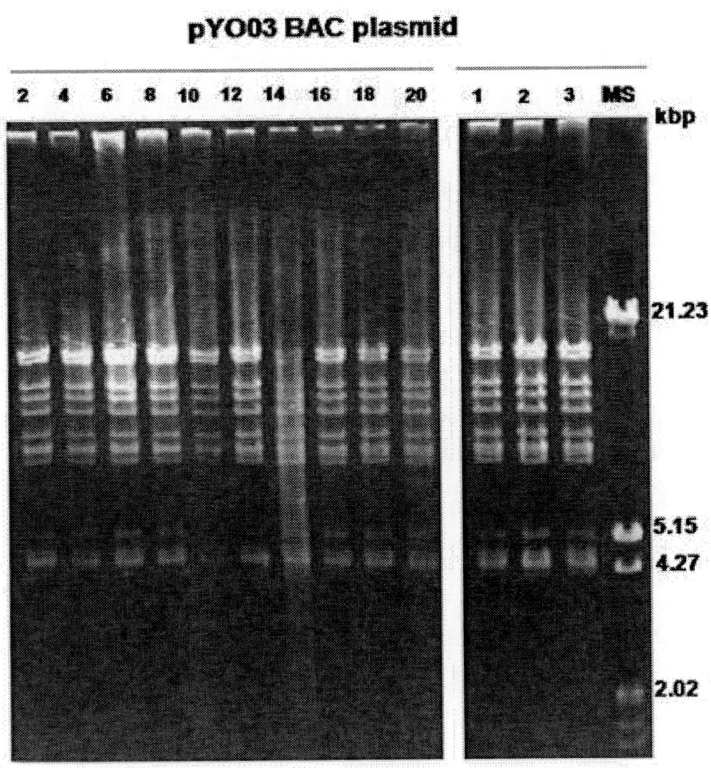

Figure 4. Stability of the EHV-4 BAC plasmid (pYO03) in *E. coli*. DH10B cells containing pYO03 were passaged every day for 20 consecutive days. At the indicated days, BAC DNAs were prepared from the culture and digested with *Bam*HI (left panel). Afterwards, bacteria containing pYO03, obtained from the last passage, were plated on agar containing chloramphenicol. DNA was isolated from single colonies and digested with *Bam*HI (lanes 1, 2, and 3; right panel). Marker sizes (MS) in kbp are indicated on the right.

IV. DNA ENGINEERING IN *E. COLI*

Cloning the herpesvirus genomes as infectious BACs allows shuttling of the genomes between mammalian cells and *E. coli*. These viral genomes can be maintained, propagated and mutated in *E. coli*. Finally, at the very end of all manipulations of the BAC DNA, it can be transferred into permissive mammalian cells to reconstitute recombinant virus. Making these large genomes in *E. coli* is only the beginning of the manipulations and functional analyses of their genes. Precise generation of recombinant DNA molecules by standard genetic engineering techniques with restriction enzymes and ligases is lost when dealing with large DNA molecules which have many restriction sites even the rare ones. Additionally, large DNA molecules are prone to breakage when dealing with them *in vitro*. Homologous recombination presents an alternative way to modify large DNA in *E. coli* through the use of bacterial recombination machinery. It allows the exchange of genetic information between two DNA molecules sharing homology regions in a precise and specific manner. These qualities are optimal for engineering a DNA molecule regardless of its size. Because the sequence of the homology regions can be chosen freely, any position on the target DNA can be specifically modified.

In *E. coli*, homologous recombination mechanisms initiated by the bacterial RecA and RecBCD enzymes impedes the use of linear DNA fragments, probably due to the aggressive exonuclease activity of RecBCD enzyme [39, 175, 176]. Therefore, herpesvirus BACs require bacterial strains in which the *E. coli* RecABCD recombination system is disrupted. In many laboratory strains of *E. coli*, the RecA gene is inactivated to increase the stability of cloned sequences. For mutagenesis, the RecA function has to be provided transiently to allow allelic exchange by homologous recombination to proceed. This can be achieved by temporarily shuttling the viral BAC to a $RecA^{+}$ bacterial strain [84, 173], by using a bacterial strain in which RecA is conditionally expressed [102, 125, 199], or by introducing a functional RecA gene temporarily on a suicide plasmid [26, 208].

Besides RecA-mediated mutagenesis, another recombinogenic engineering strategy was developed whereby recombination is mediated by phage-derived protein pairs, either RecE/RecT from the Rac phage (RecET cloning) or Redα/Redβ from λ phage (Red recombination) [132, 212]. The phage systems have a unique advantage in that they can catalyze efficient recombination with very short regions of sequence homology (around 50 bp) of a linear DNA fragment [212].

A. RecA-Dependent Recombination

In Rec-A dependent strategy, the Rec-A protein binds to single-strand segments of DNA, forming DNA-protein filaments that have the ability to search other DNA molecules for sequence homology. Once homology is found, the Rec-A filament pairs and can exchange strands with homologous segment, resulting in mutation. The major advantages of Rec-A dependent strategy are the inclusion of a temperature-sensitive (ts) plasmid origin and that there are no traces of foreign sequences, such as selection markers, remain in the BAC genome after mutagenesis.

Only shuttle plasmid can be used to allow the neat introduction of any kind of mutation (e.g. deletion, insertion, point mutation, or sequence replacement). Before the first round of recombinogenic engineering, a suicide plasmid, containing the desired mutation flanked by viral homologous regions to the target site within the viral BAC, must be constructed. The constructed shuttle plasmid is then transformed into the BAC-harboring *E. coli*. Allelic exchange then proceeds through a two-step process of homologous recombination events. In the first step, the shuttle plasmid is inserted into the BAC plasmid by homologous recombination, resulting in a cointegrate. In the second step, the cointegrate may spontaneously resolve to either wild type or mutant BAC plasmid (Figure 5). After resolution of the cointegrate, the shuttle plasmid will be lost by activating its temperature-sensitive origin. To improve the yield of the intended product in the second round, a counterselectable gene has to be included near the temperature-sensitive origin. Furthermore, using homologous sequences of approximately the same length increases the yield of mutants. Several herpesviral mutants have been generated using this method [6-8, 28, 157, 199, 200]. However, this method is more laborious and time consuming than other methods, since it requires construction of a shuttle plasmid via multiple cloning steps.

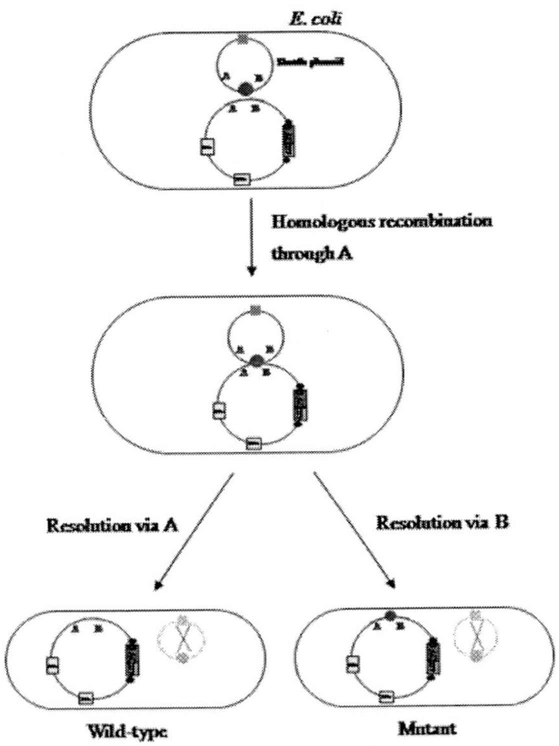

Figure 5. Site-directed mutagenesis of BACs using RecA-dependent recombination. A temperature-sensitive shuttle plasmid containing the desired mutation (violet) flanked by homologous sequences (A and B) is transformed into *E. coli* carrying the viral BAC. RecA-mediated homologous recombination through one homology arm leads to formation of a cointegrate, which is selected by the antibiotic-resistance gene (blue box) that is encoded on the shuttle plasmid. In the second round, the cointegrate is resolved to generate either a mutant BAC (right) or a wild-type viral BAC (left). The free shuttle plasmid is lost at the non-permissive temperature.

B. Red Recombination (Et Cloning)

An alternative strategy of mutagenesis was developed where linear DNA fragments can be used to introduce a mutation (Figure 6). This strategy is not dependent on RecA, but instead it uses the recombination functions RecE and RecT or Redα and Redβ for introduction of mutation into a circular DNA molecule. Many laboratories adapted this method to introduce mutations into herpesviral BACs [1, 9, 13, 22, 99, 155, 161, 190]. A linear DNA fragment containing a selectable marker and short homologous sequences flanking the target sequence is transformed into the BAC-carrying *E. coli*. By a double cross-over event, the selectable marker is introduced along with the mutation. To secure the process of recombination, the bacterial RecBCD exonuclease activity must be inactivated either by using exonuclease-negative bacteria or by expressing the exonuclease inhibitor activity of Redγ from bacteriophage λ.

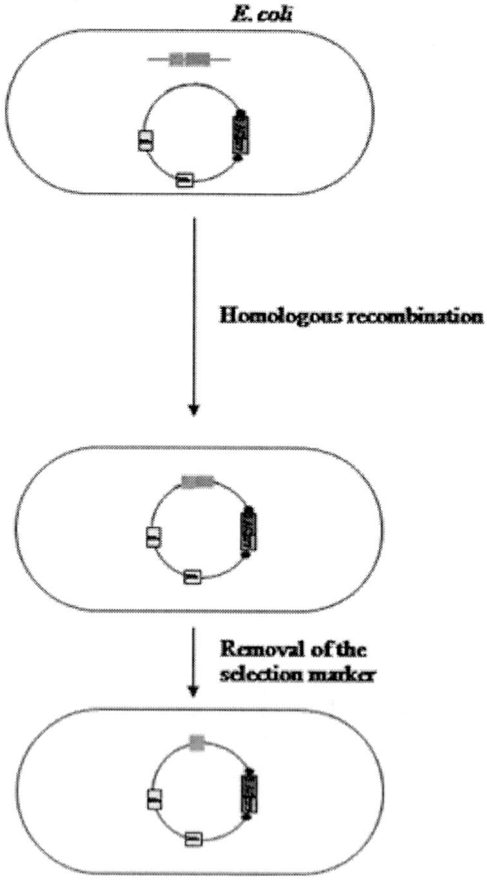

Figure 6. Site-directed mutagenesis of BACs using Red recombination. A linear DNA fragment containing the desired mutation (blue box) and antibiotic-resistance gene (orange box) flanked by homologous arms of about 50 nucleotides is transformed into *E. coli* carrying the viral BAC. The mutant genomes are identified by antibiotic selection. The selectable marker can be finally excised using the suitable recombination system.

A modified recombination system has been developed for efficient chromosome engineering in RecA$^+$ or RecA$^-$ *E. coli* by using electroporated linear DNA [210]. In this system, an *E. coli* strain was made that contains a λ prophage harboring the recombination genes *exo*, *bet*, and *gam* under the tight control of the temperature-sensitive λ (allele cI857) repressor. The *exo*, *bet*, and *gam* genes can be easily activated or inactivated by shifting the cultures to 42 ° C or 32 ° C, respectively. When λ functions are activated for as short a time as 15 min, *E. coli* becomes more recombinogenic and take up linear DNA without its destruction [131, 212]. This recombination system does not require RecA function and depends primarily on the expression of *exo*, *bet*, and *gam*. Gam inhibits the RecBCD nuclease from attacking linear DNA, and Exo and Beta generate recombination activity. More importantly, this recombination is proficient with DAN homologies as short as 25-50 nucleotides on the ends of the linear DNA. These sequences can be synthesized using PCR primer pairs that contained 20 bp specific for the selectable marker, and an additional 50 bp homologous to sequences flanking the region to be mutated [5, 82, 136]. Thus, the linear targeting DNA can be a PCR product and no dedicated recombination plasmid needs to be constructed. The selectable marker can be chosen so that it conveys resistance to antibiotics in both *E. coli* and mammalian cells. Different genes are useful, especially those that convey resistance to kanamycin/neomycin, hygromycin and blasticidin [133]. However, any other selectable genes can be used.

Although this method might appear to be more straightforward, it is also more difficult to control and it has some drawbacks. First is the potential instability of the viral BAC since short repeats in the herpesviral genome can lead to unwanted recombination. Second is the need to insert a selectable marker. Although the selectable marker can be removed by using flanking FLP or Cre target recognition sites (FRT or *loxP* sites, respectively), usually foreign sequence traces, only 34 nucleotides, are remained in the mutation site that may interfere with gene expression. For this reason, several alternative recombination systems have been developed for markerless mutagenesis of large DNA sequences in *E. coli*. One of these systems depends mainly on the selection and counterselection strategy. First, a cassette consisting of either positive and negative selection markers or bifunctional markers has to be inserted into the DNA. This allows both positive and negative selection by one gene product. Then, those cassettes are replaced with the desired markerless linear DNA by a second recombination event and subsequent counterselection [92, 202, 206]. However, one problem associated with this system is that the selection markers can be lost unspecifically due to recombination events between repeat sequences, which are common in both mammalian and viral sequences [190]. Other methods for markerless replacement without selection markers include, the use of single oligonucleotides harboring modified sequences [61], the gene gorging technique [81], or the hit-and-fix protocol [209]. However, these methods lack a powerful procedure to screen for desired recombination events.

Recently, highly efficient, PCR-based DNA engineering strategy has been developed [190]. In this modified strategy, the positive selection markers that were used to introduce the desired target mutation in a first step are removed in a second step by the combination of I-*Sce*I cleavage and intramolecular Red recombination utilizing previously introduced sequence duplication. This modification renders the method versatile, very efficient, and there is no retention of unwanted foreign sequences at the site of mutation.

C. Transposon Mutagenesis

In bacterial genetics, the use of transposons (Tns) for random mutagenesis has a long and successful tradition [18]. Tns are mobile genetic elements that can randomly insert themselves into a DNA molecule. Thus, after transfer of a Tn donor plasmid into *E. coli* already containing the BAC plasmid, the transposable element can jump into the viral BAC DNA (Figure 7). If the Tn was inserted into an ORF, this can truncate or functionally inactivate the gene. Tns have different properties; for example, some Tns prefer to insert into the bacterial genome, others into negatively supercoiled plasmids. The use of a Tn with a preference for plasmids, can allow direct isolation of *E. coli* clones with mutated BAC [26]. Otherwise, additional steps are needed to isolate the relatively infrequent Tn insertions into the BAC [173]. One advantage of Tn mutagenesis is that the molecular analysis of mutant genomes is straight-forward, as the Tn insertion sited can be directly determined by sequencing from primer sites within the Tn. Mutants with Tn insertions within specific genes can be identified by using a PCR-based screening procedure [26, 83]. Once identified, the mutation can be repaired by allelic exchange to exclude the influence of second-site mutations.

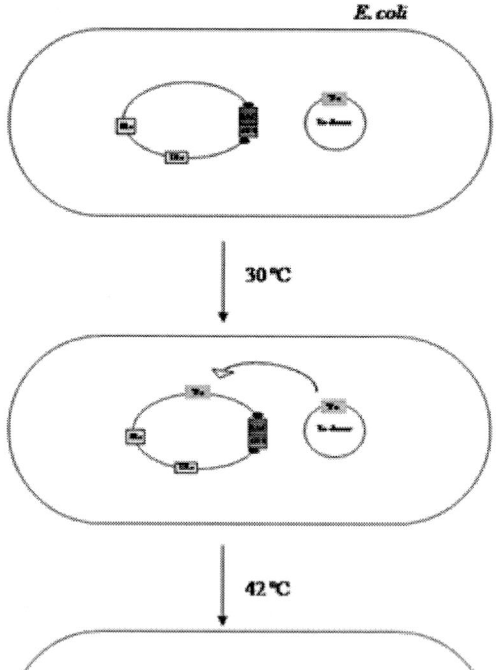

Figure 7. Transposon mutagenesis of herpesvirus BACs. A temperature-sensitive transposon donor plasmid (Tn donor) is transformed into *E. coli* carrying the viral BAC. At the permissive temperature, the Tn inserts into the BAC DNA. Appropriate selection at the non-permissive temperature removes the Tn donor plasmid and selects the Tn-inserted BAC.

Random Tn mutagenesis can be used to generate large libraries of mutant BAC genomes, which can be analyzed with phenotypic screens [29, 83]. Using such informative phenotypic screens, gene functions can be determined in a classical, but extremely rapid and efficient, genetic approach from phenotype-to-genotype. A unique approach to the classical concept of phenotype-to-genotype screening is the use of invasive bacterial as a tool for the transfer of herpesvirus BACs into mammalian cells. *E. coli* expressing certain bacterial genes, invasin and listeriolysin, can invade permissive eukaryotic cells and release the viral BAC.

V. Mutagenesis of EHV-4 Genome

The feasibility to introduce targeted mutations into the BAC cloned EHV-4 genome was demonstrated by deleting the genes encoding TK and gC by Red mutagenesis in *E. coli* [12, 13]. However, the deletion of some other genes has been attempted, such as the deletion of gG (Azab *et al.*, unpublished data). In this chapter, we will describe the construction and characterization of TK-, gC- and gG- deleted mutants.

To generate mutants of EHV-4, transfer constructs were generated by PCR amplification of the I-*Sce*I-*aphAI* cassette, from pEPkan-S plasmid, with primer pairs contained an approximately 50 bp DNA duplication of nucleotides homologous to sequences flanking the region of interest. Then, the mutant BAC plasmid was isolated from *E. coli* cultures using standard alkaline lysis procedures [156], and analyzed by restriction enzyme digestion and gel electrophoresis. Further analysis is usually done by Southern blot analysis and/or sequencing of the targeted region. To produce the recombinant virus from EHV-4 BAC plasmid, 293 cells were transfected with 2 μg of BAC DNA. Three days later, the supernatant and cells were collected and used to infect confluent FHK cells. Viral plaques were observed 2 or 3 days after infection.

An advantage of this technology, when compared with conventional mutagenesis methods in cell culture, is increased safety. Since the viral genome is kept as a BAC throughout all steps of manipulation and there is no need to handle infectious viruses. Additionally, the viral sequences at the time of cloning are preserved, since *in vitro* cell culture of the virus is not necessary.

A. Construction and Characterization of a Thymidine Kinase-Deficient Mutant of EHV-4

The study of TK-deficient mutants of other herpesviruses has received considerable attention for many years. Deletion of this viral encoded enzyme in alphaherpesviruses, such as HSV-1, HSV-2, BHV-1, and PrV, resulted in a reduction in virulence and the rate of re-activation of latent viruses [34, 64, 103, 128, 169, 182]. Previous reports of EHV-1, the close relative of EHV-4, TK-deficient mutants suggested that they are markedly less pathogenic than the wild-type virus, despite being able to replicate in the host [38, 169]. More recently, the role of TK in EHV-4 pathogenesis has been elucidated [13].

To generate EHV-4ΔTK mutant, DNA of the UL23 (TK) deficient pYO03, named pYOΔTK, was obtained after electroporation of TK recombination cassette into *E. coli*

EL250 harboring EHV-4BAC (pYO03) [12], and the bacteria were plated on LB agar plates containing kanamycin and chloramphenicol. Double-resistant colonies were shown to exhibit the proper insertion of kanamycin resistant (Kan^R) gene and the deletion of TK after PCR, restriction enzyme digestion and DNA sequencing analyses. The reconstitution of mutant virus was performed by transfection of pYOΔTK into 293 cells (Figure 8). These transfected cells could be detected by expression of GFP encoded within pEH4BAC sequences [12]. After 2 days, the supernatants and cells were collected and used to infect confluent FHK cells. Viral plaques were observed 2 or 3 days after infection as well as CPE was spread into the culture. DNA of the recombinant virus and wild-type virus was prepared and subjected to PCR, Southern blot and DNA sequencing analyses. The expected genotype of the TK-mutant virus was confirmed and that Kan^R gene had been inserted instead of the TK gene.

Like other herpesviruses, EHV-4ΔTK grows normally in cell culture and has the same plaque morphology as the wild-type EHV-4 virus. Previous reports of BHV-1 [104], PrV [103, 122], and infectious laryngotracheitis virus [73] indicated that the TK-deleted recombinants were indeed less pathogenic than the wild-type viruses and they could still induce a protective immune response when inoculated in their hosts. Therefore, similarly modified EHV-4 would probably retain antigenicity and also be a candidate for a genetically engineered vaccine. However, further *in vivo* studies are required to evaluate the role of EHV-4 TK-negative mutant in immunoprophylaxis.

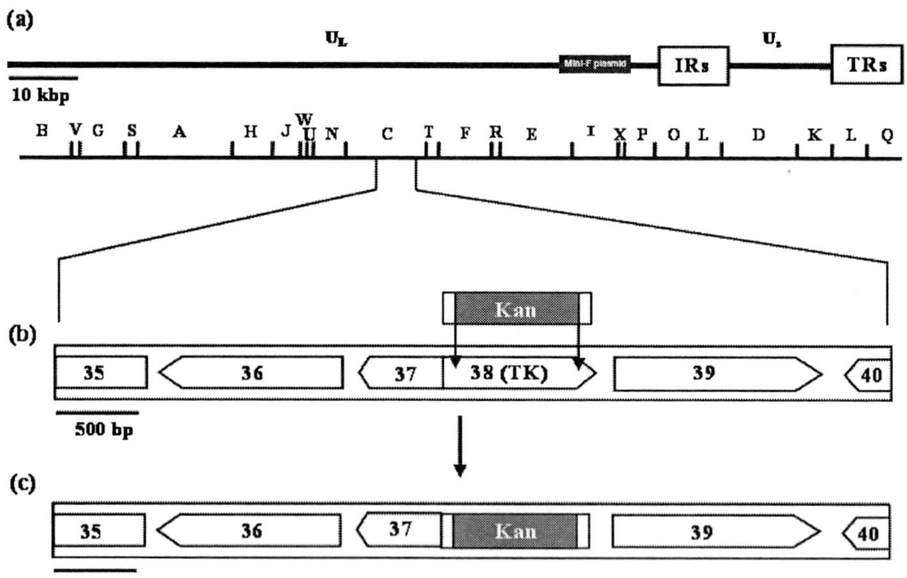

Figure 8. Schematic diagram of the procedures used to delete the TK gene from pYO03. (a) Schematic representation of the genomic organization and the *Bam*HI restriction map of EHV-4 BAC pYO03 [12] is given. The two unique regions (U_L and U_S), the terminal and internal repeat sequences (TR_S and IR_S), and the inserted BAC cassette are shown. (b) The genomic organization of the domain that encodes the genes from 35 to 40 [189] is depicted within the *Bam*HI C restriction fragment. The PCR cassette conferring Kan^R gene was inserted into the TK locus of pYO03 using red recombination. (c) The mutant BAC DNA (pYO03ΔTK) was transfected into eukaryotic cells to reconstitute the infectious virus EHV-4ΔTK.

A.1. Effect of antiviral drugs on EHV-4ΔTK

The science of antiviral chemotherapy has developed over the last 50 years. In the last two decades the rate of progress has accelerated markedly and more new chemical therapeutics have been introduced [65]. Antiviral chemotherapy is a standard practice in the management of herpesvirus infections in humans, and currently there are more than 11 licensed anti-herpetic drugs available in the human clinic. Two important groups of antiviral compounds, acyclic nucleoside analogues [e.g. acyclovir (ACV), ganciclovir (GCV) and penciclovir (PCV)] and acyclic nucleoside phosphonates [e.g. cidofovir and adefovir] are currently used in the treatment of herpesvirus infections [50, 65]. The antiviral activity and selectivity of acyclic nucleoside analogues is based on their phosphorylation through three consecutive phosphorylation steps before they can interact, in their triphosphate form, with their target enzyme, the viral DNA polymerase (Figure 9). Of crucial importance in this phosphorylation process is the first phosphorylation step which is ensured by a specific virus-encoded TK [for HSV and Varicella Zoster (VZV)] or a specific virus-encoded (UL97) protein kinase (PK) (for cytomegalovirus). Once the antiviral compounds have been phosphorylated to the monophosphate form, cellular kinases will afford their further phosphorylation to the di- and triphosphate stages. In their triphosphate form, the compounds then interact as competitive inhibitors or alternate substrates with the normal substrates [2'-deoxynucleoside 5'-triphosphates (dNTPs)], and if they are incorporated into the DNA chain, they may act as chain terminators, thus preventing further chain elongation [35, 60]. Since the acyclic nucleoside phosphonates already contain a phosphate group, they need only two, instead of three, phosphorylation steps to reach the active metabolite stage where they can interact with the DNA polymerase, resulting in inhibition of viral DNA synthesis (Figure 9). Thus, acyclic nucleoside phosphonates do not depend on the virus-induced kinase to exert their antiviral action [49, 50].

As mentioned above herpesvirus-encoded TKs have an important role in the mode of action of several acyclic nucleoside analogues (e.g. ACV and GCV) which are potent anti-herpetic drugs. Although antiviral chemotherapy is a standard practice in the management of herpesvirus infection in humans, the veterinary use of antiviral drugs is relatively uncommon. Perhaps the most frequently reported use of antiviral drugs in veterinary medicine is for the treatment of feline herpesvirus 1 infections. Nonetheless, there are some reports declared the efficacy of antiviral drugs against EHV-1, but not EHV-4, *in vitro* [51, 70, 91]. ACV and GCV were previously reported to have potent anti-EHV-1 effects by plaque reduction assay (PRA) *in vitro* [70]. Our recent work [13] was the first to determine the effect of antiviral drugs against EHV-4.

The PRA data presented in that work have confirmed the previous findings related to EHV-1. The IC_{50} (2.7 μg/ml and 0.1 μg/ml) of ACV and GCV, respectively, were approximately similar to the previous report conducted by Garre *et al.* [70]. In case of EHV-4, the obtained data have revealed that EHV-4 is relatively insensitive to ACV (IC_{50} value was >50 μg/ml); while GCV was found to exhibit a more potent specific anti-EHV-4 activity where the IC_{50} value was 2.5 μg/ml (Table 2). As antiviral compounds not only reduce the plaque number, but also contribute in the reduction of plaque size [70, 93, 196], we found that both compounds were able to significantly reduce the plaque size of EHV-4 [13]. Thus, measuring reduction in plaque number and plaque size are two valuable and complementary means of assessing the antiviral efficacy.

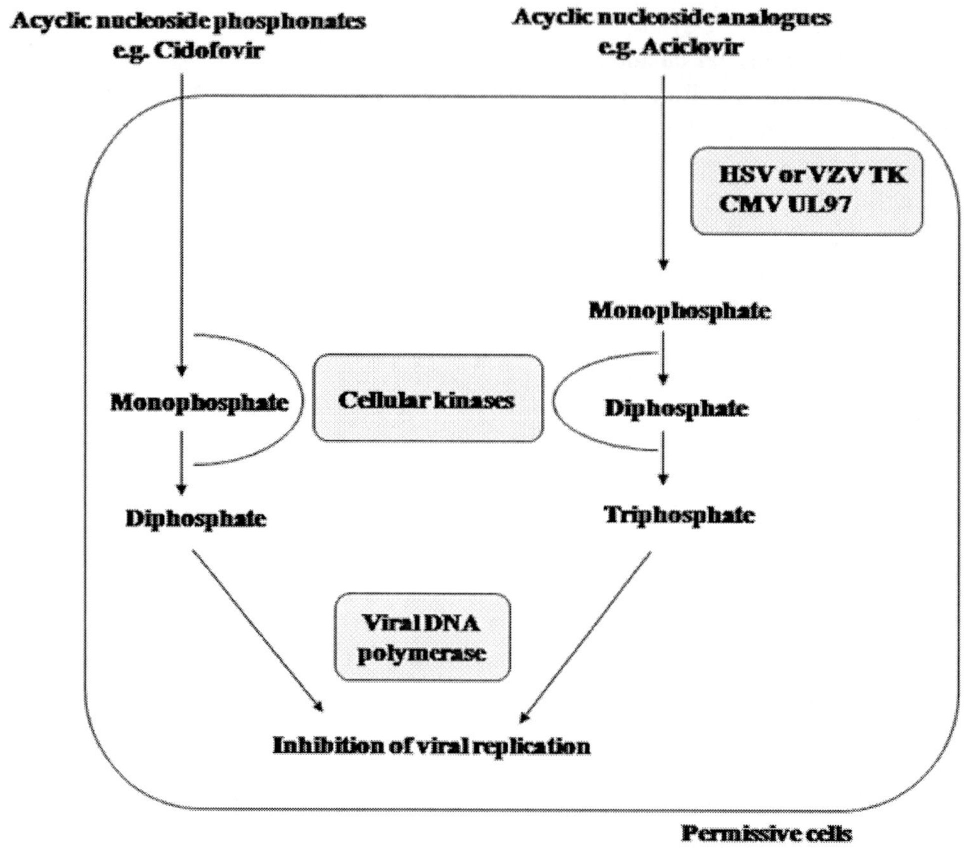

Figure 9. Mechanism of action of anti-herpes drugs targeted at the viral DNA polymerase. Nucleoside analogues are first activated by HSV- or VZV-encoded TK, or the cytomegalovirus UL97-encoded kinase, followed by phosphorylation by cellular kinases. Nucleoside phosphonates are only phosphorylated twice by cellular kinases to reach the active form.

Table 2. Susceptibility of EHV-4, EHV-1 and WA79ΔTK to antiviral drugs

Virus	Mean $IC_{50} \pm$ SD (μg/ml)	
	GCV	ACV
EHV-1	2.7 ± 0.2	0.1 ± 0.3
EHV-4	>50	2.5 ± 0.5
WA79	>50	2.1 ± 0.27
EHV-4ΔTK	>50	45
EHV-4ΔTK + pFEH4TK	>50	4 ± 0.25

Results are presented as the mean IC50 values of three independent experiments ± SD

On the other hand, EHV-4ΔTK was found to be resistant to GCV (22-fold change in IC_{50} compared to wild-type virus). When EHV-4 TK was supplied to FHK cells, GCV retained its potent inhibitory effect against the TK-deficient virus (Table 2) [13]. These results are in agreement with the nucleoside analogues resistance phenotype previously reported in other studies [23, 91, 130, 162]. GCV as well as other nucleoside analogues all rely on virus-encoded nucleoside kinases, e.g TK of HSV or EHV, or protein kinase of cytomegalovirus, for their initial phosphorylation. Therefore, TK-mutant viruses are generally insensitive to these drugs.

A.2. Antiviral susceptibility of EHV-4 in EHV-1 TK-transformed cells

The obtained PRA results showed that the IC_{50} values of ACV for EHV-1 and EHV-4 were 2.7 μg/ml and >50 μg/ml, respectively [13]. The relative insensitivity of EHV-4 to ACV could be attributed to either the inefficient phosphorylation of ACV to its 5'-monophosphate form by the EHV-4 TK or the inefficient inhibition of the viral DNA polymerase by the active form of the drug. EHV-1 TK is known to phosphorylate ACV albeit at a relatively low efficacy than that of HSV-TK. However, the *in vitro* susceptibility of EHV-1 to ACV and the anecdotal effectiveness of therapy, make it an attractive candidate for the antiviral therapy against EHV-1 infection in horses [70, 153, 187, 204]. The marked reduction in the IC_{50} value of ACV (Table 3) against EHV-4 when EHV-1 TK was supplied in trans provided direct evidence that EHV-4 was sensitive to the active tri-phosphate form of ACV. Therefore, the inefficient phosphorylation of ACV by EHV-4 TK was responsible for the observed *in vitro* insensitivity of this virus to the antiviral drug action of ACV. It is noteworthy that there is a high degree of conservation, more than 90 %, between both EHV-4 and 1 TK genes. These results clearly indicated that, despite the high sequence and structural similarity between EHV-4 TK and its EHV-1 homologue, ACV is more efficiently phosphorylated by EHV-1 TK than by EHV-4 TK. Furthermore, it could be assumed that EHV-4 DNA polymerase has a high affinity for the ACV triphosphates, and, hence explains the observed inhibitory effect on virus replication.

B. Construction and Characterization of A gC-Negative Mutant of EHV-4

Many alphaherpesviruses, such as HSV-1 [78, 207], VZV [216], PrV [126], BHV-1 [140], and EHV-1 [141] attach to target cells by interaction of the virion with the cell-surface heparan sulfate. In HSV-1, PrV, BHV-1 and EHV-1, the viral gC is primarily responsible for this interaction. Evidence for this includes the following: (i) these viruses exhibit greatly reduced binding to cell surfaces devoid of heparan sulfate, regardless of whether this deficiency is due to enzymatic treatment with heparinase or heparitinase or mutant cells that fail to synthesize heparan sulfate [72, 80, 126, 140, 141, 165, 207]. (ii) Heparin, which is structurally and chemically very similar to heparan sulfate [110], and heparin binding proteins (e.g. platelet factor 4 and neomycin) inhibit the binding of virions to cells [78, 79, 112, 140, 207]. (iii) Deletion of the gC coding genes of HSV-1, PrV, BHV-1 or EHV-1 results in virions that are impaired in their ability to bind to cells [108, 141, 160]. In contrast to these viruses, HSV-2 gC does not play the key role in viral binding, however, HSV-2 gC is a heparin-binding glycoprotein [71]. Interestingly, HSV-1 gB is also a heparin-binding glycoprotein and does participate in the binding process to cell surface heparan sulfate, at least when gC is absent from the virion envelope [71, 80]. Although gC plays the predominant role in viral binding during the early steps of infection, gC is dispensable for virus growth and gC-negative mutants are still infectious, but exhibit decreased levels of binding and infectivity compared with wild-type virus [78, 181].

Table 3. Comparative antiviral susceptibility of EHV-4 in pFEH1TK versus negative control pFLAG-CMV-2 transfected cells by PRA

Drug	Mean IC50 ± SD (μg/ml)	
	pFLAG-CMV-2	pFEH1TK
ACV	>50	2.9 ± 0.26
GCV	2.7 ± 0.5	0.5 ± 0.3

Results are presented as the mean IC_{50} values of three independent experiments ± SD

For EHV-4, the contribution of individual glycoproteins to viral binding has not been experimentally detected. It may be assumed that the EHV-4 gC performs functions in viral binding similar to that of the gC homologue of its close relative EHV-1. However, although EHV-1 and EHV-4 are closely related and they show a high degree of conservation ranging between 55 %- 96 % similarity between individual genes [189], the two viruses exhibit differences in epidemiology, pathogenicity and cell tropism that may reflect differences in viral entry. More recently, the role of EHV-4 gC in viral binding has been identified [14].

The gC-negative mutant, EHV-4ΔgC, was constructed using a construction approach similar to that used for the generation of TK-negative mutant (Figure 10) [13]. PCR analysis showed that primers binding to the outside of the deleted part of gC yielded a PCR product of around 1.6 kbp and 1 kbp in case of wild-type EHV-4 and EHV-4ΔgC, respectively (Figure 11a). Southern blot analysis (Figure 11b) revealed that the mutants lacking the gC gene failed to react with the gC probe, while the wild-type virus reacted with the gC probe. These results and those obtained by DNA sequencing confirmed the correct insertion of the Kan^R gene instead of the gC gene. Like other alphaherpesviruses, *in vitro* analysis revealed that gC is dispensable for viral propagation in cell culture and the gC-negative mutant exhibited slightly lowered virus titers when compared to parental viruses.

B.1. The role of EHV-4 gC in viral binding to cell surface heparan sulfate

Like several other herpesviruses, EHV-4 binds to host cells through the interaction of the virus with a cell surface component that includes a heparinlike moiety. This conclusion emerges from the finding that agents which can block this interaction effectively block EHV-4 infection. Soluble heparin was capable of preventing infection of FHK cells when EHV-4 virions were preincubated with heparin. Presumably heparin occupies the sites necessary for attachment of virions to cell surface heparan sulfate. Enzymatic removal of heparan sulfate from the cell surface significantly reduced the infection of cells by EHV-4. Furthermore, suppression of GAG synthesis by addition of sodium chlorate interfered with EHV-4 plaque formation in FHK cells. Sodium chlorate treatment of cells is known to inhibit heparan sulfate-dependent infection of different viruses through the inhibition of sulfation of various GAG and proteins without inhibiting cell growth or protein synthesis [33, 101, 146]. These results strongly suggested that heparan sulfate either is required for infectivity or greatly enhances the efficiency of infection. Inasmuch as gC has heparin-binding activity, it seems likely that gC confers binding of the virion to cell surface heparan sulfate (Figure 12). However, EHV-4 is also able to use receptors and entry pathways other than those mediated by viral gC.

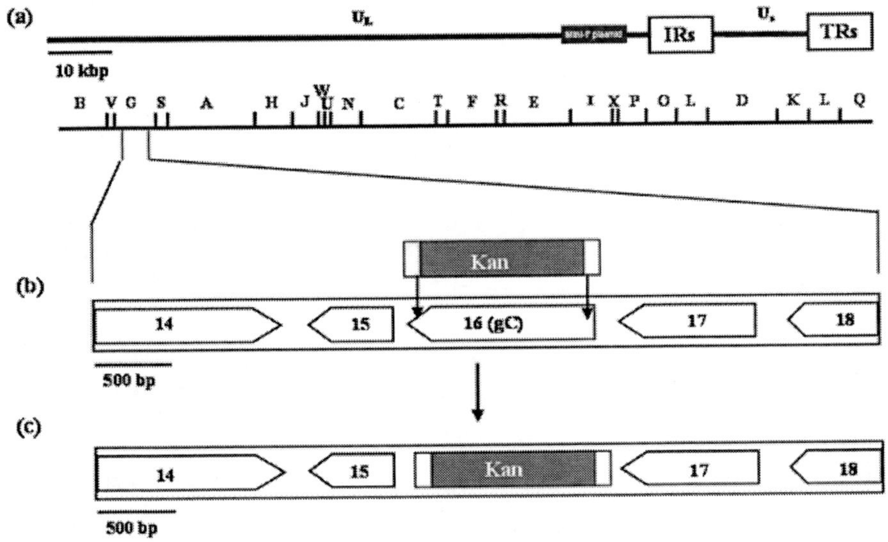

Figure 10. Schematic diagram of the procedures used to delete the gC gene from pYO03. (a) Schematic representation of the genomic organization and the *Bam*HI restriction map of EHV-4 BAC pYO03 [12] is given. The two unique regions (U_L and U_S), the terminal and internal repeat sequences (TR_S and IR_S), and the inserted BAC cassette are shown. (b) The genomic organization of the domain that encodes the genes from 14 to 18 [189] is depicted within the *Bam*HI G restriction fragment. The PCR cassette conferring Kan^R gene was inserted into the gC locus of pYO03 using red recombination. (c) The mutant BAC DNA (pYO03ΔgC) was transfected into eukaryotic cells to reconstitute the infectious virus EHV-4ΔgC.

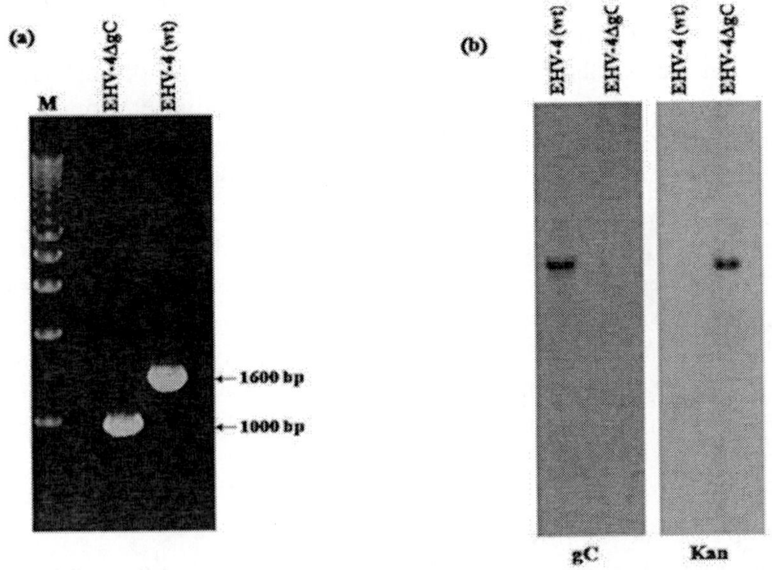

Figure 11. Identification of EHV-4ΔgC by a combined PCR/Southern blot analysis. (a) PCR analysis of EHV-4 mutants to detect gene 16 encoding gC and the Kan^R gene. PCR products from wild type EHV-4 and mutant virus EHV-4ΔgC were electrophoresed in a 1 % agarose gel. Marker (lane M) was included on the left. (b) Purified DNAs from the viruses digested with *Bam*HI and tested by Southern blotting using probes corresponding to gene 16 of the wild type EHV-4 (left panel) and Kan^R gene (right panel).

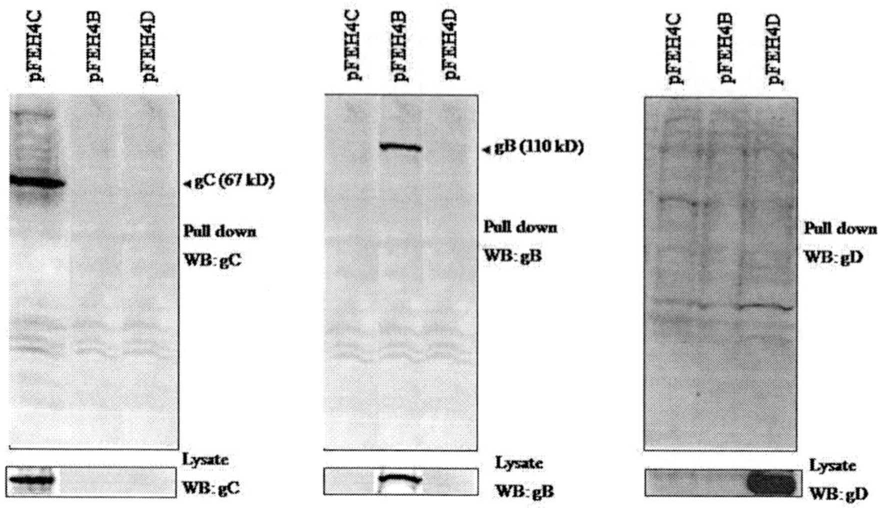

Figure 12. Binding of EHV-4 glycoproteins to heparin-agarose beads. 293 cells were transfected with pFEH4C, pFEH4B or pFEH4D, harvested and lysed with lysis buffer. The cell lysates were then mixed with heparin-agarose beads. The proteins that had bound to the beads were analyzed by Western blotting using anti-gC (left panel), anti-gB (middle panel), or anti-gD (right panel) polyclonal antibodies. Arrowheads show the specific bands of gC and gB.

The role of gC in virus attachment and penetration is well documented for both HSV-1 and PrV [71, 78-80, 126, 207]. In both viruses, gC plays the predominant role in mediating the binding of the virion to cell surface heparan sulfate. However, mutants devoid of gC are still infectious, but exhibit decreased levels of binding and infectivity compared with wild-type viruses [78, 97, 98]. In HSV-1 and BHV-1, another glycoprotein, gB, also binds to heparan sulfate and plays the key role in the attachment of gC-deleted viruses to the cell surface proteoglycans. In contrast, the binding of gC-negative PrV mutant to cells is not dependent on heparan sulfate [80, 105, 113, 126]. The studies described in our recent report demonstrate that EHV-4 gC also confers the binding of the virion to cell surface heparan sulfate and gC-negative EHV-4 is still dependent on interactions with heparan sulfate for virus entry (Figure 13) [14]. There must be at least another viral glycoprotein that can mediate the interaction with heparan sulfate moieties. Pull down assay studies indicated that EHV-4 gB is also a heparin-binding glycoprotein. Furthermore, a recent study has shown that gB-mutated EHV-4 is more susceptible to heparin than wild-type virus (K. Andoh, K. Kai, T. Matsumura, and K. Maeda, unpublished results). Thus, we would suggest that EHV-4 gC and gB have roles in the initial events of infection and could mediate the adsorption of virus to cells.

B.2. Glycoprotein C has no role in virus penetration

A previous report has indicated that EHV-1 gC-negative viruses were severely impaired in the rate of penetration [141]. However, our recently obtained results revealed that there was no delay in the penetration of adsorbed EHV-4ΔgC virus compared to that of the wild-type EHV-4 (Figure 14). Whereas the half time of penetration of EHV-1 gC-deleted virus was greater than 120 min, it was only around 30 min for EHV-1 [141], EHV-4 and EHV-4ΔgC [14].

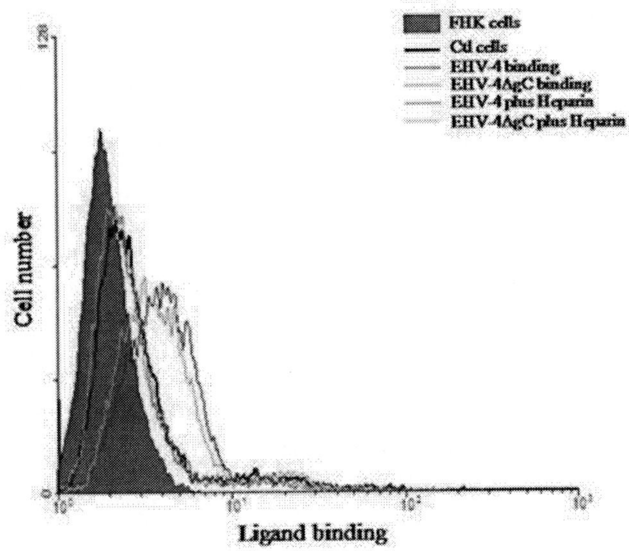

Figure 13. Binding of EHV-4 and EHV-4ΔgC to FHK cells. FHK cells were incubated with EHV-4 or EHV-4ΔgC in the presence or absence of heparin (1000 μg/ml). Cell surface binding was detected by flow cytometry. The histogram shown is representative of two independent experiments. Ctl cells, control cells incubated with the primary and secondary antibodies without prior virus binding.

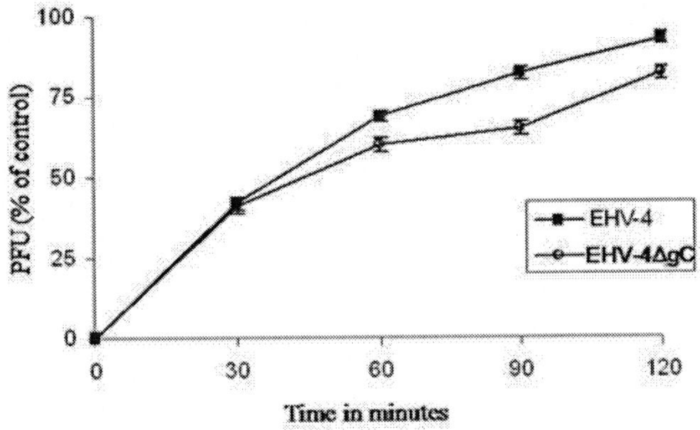

Figure 14. Penetration assay of EHV-4 and EHV-4ΔgC. The rate of penetration on FHK cells was determined as the percentage of the plaques produced at the given time points after citrate treatment in relation to untreated cells. The values represent the mean ± SD of two independent experiments.

Based on these results, we can conclude that the relative contributions of gC for viral attachment as well as post-binding events for EHV-4 and EHV-1 may be quite different although there is a high sequence and structural similarity between EHV-4 gC and its EHV-1 homologue. These differences may help explain the difference in cell tropism between both viruses. However, previous reports have shown that the differences in cell tropism between EHV-1 and EHV-4 are partly due to the differences in their gD and gG homologues [193, 203]. Taken together, one may speculate that difference in cell tropism between EHV-1 and EHV-4 may well be associated with differences in the contributions of different glycoproteins, at least gC, gD and gG, which involved in virus entry and/or spread from an infected to a neighboring uninfected cell.

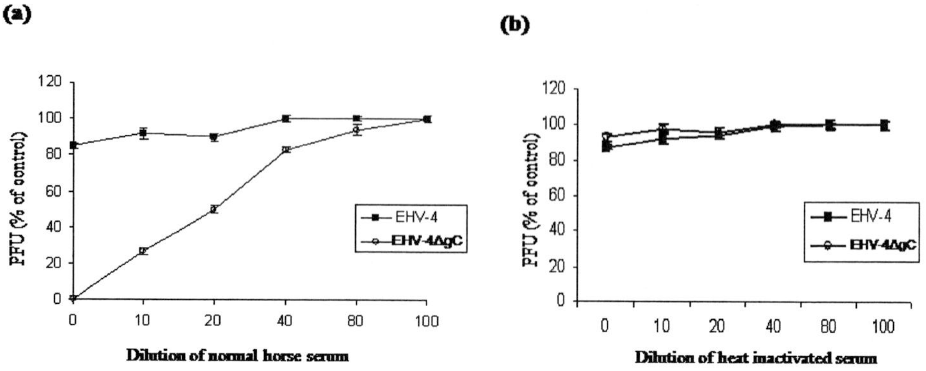

Figure 15. Effects of normal horse serum (complement) on EHV-4 infectivity. Normal (a) or heat inactivated horse serum (b), free of anti-EHV-4 and EHV-1 antibody, were mixed in the proportions indicated with different viruses. Infectious titers were determined on FHK cells. The values presented are the mean of two independent experiments.

B.3. Glycoprotein C protects EHV-4 from complement-mediated neutralization

Previous studies indicated that beside the heparin-binding activity of HSV-1 and HSV-2 gC, it also binds to the C3b fragment of the third component of complement *in vitro* [59, 67, 89, 123]. It has been shown that HSV-1 gC can accelerate the decay of the C3 convertase of the alternative pathway, whereas HSV-2 gC stabilizes this enzyme [68, 76, 88]. This activity has been proven to contribute for the protection of HSV-1 from the host immune system and thus gC-C3 interactions are involved in viral pathogenesis [111]. Likewise, a direct interaction between EHV-4 gC and equine C3 has been demonstrated and this interaction exhibited species-specificity [89, 90]. However, the importance of gC in protecting EHV-4 from complement-mediated neutralization had not been tested. In our study, we found that gC also protects EHV-4 from complement-mediated lysis (Figure 15). Whether this C3-binding activity contributes to the pathogenesis and immune evasion of EHV-4 *in vivo* remains to be determined.

C. *In Vitro* Characterization of EHV-4 gG-Deleted Mutant

Glycoprotein G homologues have been identified in most alphaherpesviruses [36, 40, 55, 62, 63, 69, 119, 148, 171, 177]. gG is unusual among herpesvirus glycoproteins in that it is cleaved after translation and the larger N terminal cleavage product is secreted into the medium. In addition to the secreted form, gG of EHV-4 and EHV-1 has been shown to have two additional virion-associated forms [54]. However, the only exception is HSV-1 gG, which apparently lacks a secreted form [40, 54, 62, 100]. Although gG has been shown to be nonessential for virus replication *in vitro*, the exact role of gG in herpesvirus infection is not fully understood. In contrast to other herpesvirus glycoproteins, which are under the transcriptional control of late promoters, gG expression appears to be under the control of early promoters [37, 62, 100].

Recently, gG homologues of several alphaherpesviruses have been shown to function as a viral chemokine-binding protein (vCKBP) that binds a broad range of chemokines in a species-independent fashion. gG binds to chemokines with high affinity and inhibits their biological activity *in vitro* by preventing their interaction specific cell receptors [30]. As long as chemokines play central roles in mediating inflammatory responses during viral infections, the chemokine binding properties of gG suggest that it might be involved in virus immune evasion. Experimental studies with EHV-1 gG indicated that gG not only binds to chemokines but is also capable of inhibiting their chemotactic function both *in vitro* and *in vivo*, thereby contributing to viral pathogenesis and virulence [193, 195]. On the other hand, EHV-4 gG does not exhibit properties consistent with a vCKBP interfering with at least IL-8-induced chemotaxis [193].

Deletion of EHV-4 gG has been previously reported [86]. *In vitro* characterization of this mutant indicated that EHV-4ΔgG grew more slowly in cell culture and produced plaques of different morphology. This could be attributed to the unexpected deletions in the adjacent downstream gene ORF 71. However, after the construction of EHV-4 BAC clone, the process of mutagenesis becomes precise and more accurate. In our case, we could delete EHV-4 gG without any deletions of any of the surrounding genes. Our data showed that the mutant virus exhibited comparable phenotypes to that of the wild-type virus. The growth rates, titers, as well as plaque sizes of the wild-type EHV-4 and gG-mutant viruses were virtually indistinguishable (Azab *et al.*, unpublished data).

VI. Conclusion and Future Prospects

EHV-4 is a major cause of respiratory tract disease in horses worldwide. Efficient generation of mutants of EHV-4 would significantly contribute to the rapid and accurate characterization of the viral genes. The construction of EHV-4 BAC clone will be an invaluable resource for the production of recombinant viruses. This system will be a very useful tool for analyzing the function of different genes of this important equine viral pathogen and may allow the production of recombinant viral vaccines to control EHV-4 and EHV-1. EHV-4 possesses two main properties which make it a good candidate for the development of recombinant vaccines for horses; first, in contrast to EHV-1, the clinical disease caused by EHV-4 is milder than that of EHV-1. Second, the immunity developed against the recombinant vaccine should confer cross-protection against EHV-1.

Despite the high sequence and structural similarity, more than 90 %, between EHV-4 TK and its EHV-1 homologue, ACV was more efficiently phosphorylated by EHV-1 TK than by EHV-4 TK. This explains why ACV is an attractive candidate as antiviral drug against EHV-1, but not EHV-4, infection in horses. However, GCV displayed a potent activity against both EHV-4 and 1 infection *in vitro*. But there is no direct clinical application due to its high price cost.

The role of EHV-4 gC in the adsorption of the virus to cell-surface heparan sulfate has been elucidated. However, gB, another heparin-binding glycoprotein, could also mediate the adsorption of EHV-4 to the cells. Moreover, gC may also play roles in the protection of EHV-4 from the complement-mediated neutralization. Again, another difference between EHV-4 and its close relative EHV-1 has been shown. This was clear from the findings that the relative contributions of gC for viral attachment and penetration events are quite different although there is a high sequence and structural similarity between EHV-4 gC and its EHV-1 homologue.

Cloning of EHV-4 as infectious BAC has greatly speeded up the construction of mutants, thereby allowing assessment of the role of viral genes in host-virus interaction. Furthermore, the construction of EHV-4 BAC will have the potential to allow further study of aspects of the difference between EHV-4 and EHV-1.

References

[1] Adler, H., Messerle, M., Wagner, M. & Koszinowski, U. H. (2000). Cloning and mutagenesis of the murine gammaherpesvirus 68 genome as an infectious bacterial artificial chromosome. *J Virol*, *74*, 6964-74.

[2] Adler, H., Messerle, M. & Koszinowski, U. H. (2001). Virus reconstituted from infectious bacterial artificial chromosome (BAC)-cloned murine gammaherpesvirus 68 acquires wild-type properties in vivo only after excision of BAC vector sequences. *J Virol*, *75*, 5692-6.

[3] Adler, H., Messerle, M. & Koszinowski, U. H. (2003). Cloning of herpesviral genomes as bacterial artificial chromosomes. *Rev Med Virol*, *13*, 111-21.

[4] Allen, G. P. & Bryans, J. T. (1986). Molecular epizootiology, pathogenesis, and prophylaxis of equine herpesvirus-1 infections. Prog Vet *Microbiol Immunol*, *2*, 78-144.

[5] Angrand, P. O., Daigle, N., van der Hoeven, F., Scholer, H. R. & Stewart, A. F. (1999). Simplified generation of targeting constructs using ET recombination. *Nucleic Acids Res, 27*, e16.

[6] Angulo, A., Messerle, M., Koszinowski, U. H. & Ghazal, P. (1998). Enhancer requirement for murine cytomegalovirus growth and genetic complementation by the human cytomegalovirus enhancer. *J Virol*, *72*, 8502-9.

[7] Angulo, A., Ghazal, P. & Messerle, M. (2000). The major immediate-early gene ie3 of mouse cytomegalovirus is essential for viral growth. *J Virol*, *74*, 11129-36.

[8] Angulo, A., Kerry, D., Huang, H., Borst, E. M., Razinsky, A., Wu, J., Hobom, U., Messerle, M. & Ghazal, P. (2000). Identification of a boundary domain adjacent to the potent human cytomegalovirus enhancer that represses transcription of the divergent UL127 promoter. *J Virol*, *74*, 2826-39.

[9] Arii, J., Hushur, O., Kato, K., Kawaguchi, Y., Tohya, Y. & Akashi, H. (2006). Construction of an infectious clone of canine herpesvirus genome as a bacterial artificial chromosome. *Microbes Infect*, *8*, 1054-63.

[10] Arii, J., Kato, K., Kawaguchi, Y., Tohya, Y. & Akashi, H. (2009). Analysis of herpesvirus host specificity determinants using herpesvirus genomes as bacterial artificial chromosomes. *Microbiol Immunol*, *53*, 433-41.

[11] Avitabile, E., Lombardi, G., Gianni, T., Capri, M. & Campadelli-Fiume, G. (2004). Coexpression of UL20p and gK inhibits cell-cell fusion mediated by herpes simplex virus glycoproteins gD., gH-gL., and wild-type gB or an endocytosis-defective gB mutant and downmodulates their cell surface expression. *J Virol*, *78*, 8015-25.

[12] Azab, W., Kato, K., Arii, J., Tsujimura, K., Yamane, D., Tohya, Y., Matsumura, T. & Akashi, H. (2009). Cloning of the genome of equine herpesvirus 4 strain TH20p as an infectious bacterial artificial chromosome. *Arch Virol*, *154*, 833-42.

[13] Azab, W., Tsujimura, K., Kato, K., Arii, J., Morimoto, T., Kawaguchi, Y., Tohya, Y., Matsumura, T. & Akashi, H. (2010). Characterization of a thymidine kinase-deficient mutant of equine herpesvirus 4 and in vitro susceptibility of the virus to antiviral agents. *Antiviral Res*, *85*, 389-95.

[14] Azab, W., Tsujimura, K., Maeda, K., Kobayashi, K., Mohamed, Y. M., Kato, K., Matsumura, T. & Akashi, H. (2010). Glycoprotein C of equine herpesvirus 4 plays a role in viral binding to cell surface heparan sulfate. *Virus Res*.

[15] Azmi, M. & Field, H. J. (1993). Interactions between equine herpesvirus type 1 and equine herpesvirus type 4: T cell responses in a murine infection model. *J Gen Virol*, *74*, 2339-45.

[16] Bender, F. C., Whitbeck, J. C., Ponce de Leon, M., Lou, H., Eisenberg, R. J. & Cohen, G. H. (2003). Specific association of glycoprotein B with lipid rafts during herpes simplex virus entry. *J Virol*, *77*, 9542-52.

[17] Bender, F. C., Whitbeck, J. C., Lou, H., Cohen, G. H. & Eisenberg, R. J. (2005). Herpes simplex virus glycoprotein B binds to cell surfaces independently of heparan sulfate and blocks virus entry. *J Virol*, *79*, 11588-97.

[18] Berg, C. M., Berg, D. E. & Groisman, E. A. (1989). Transposable elements and the genetic engineering of bacteria. In *Mobile DNA* (Berg, D.E. and Howe, M.M., Editors): pp. 879-926, American Society for Microbiology, *Washington*, DC, USA

[19] Borchers, K., Wolfinger, U., Lawrenz, B., Schellenbach, A. & Ludwig, H. (1997). Equine herpesvirus 4 DNA in trigeminal ganglia of naturally infected horses detected by direct in situ PCR. *J Gen Virol*, *78*, (Pt 5), 1109-14.

[20] Borchers, K., Wolfinger, U. & Ludwig, H. (1999). Latency-associated transcripts of equine herpesvirus type 4 in trigeminal ganglia of naturally infected horses. *J Gen Virol*, *80*, (Pt 8): 2165-71.

[21] Borst, E. M., Hahn, G., Koszinowski, U. H. & Messerle, M. (1999). Cloning of the human cytomegalovirus (HCMV) genome as an infectious bacterial artificial chromosome in Escherichia coli: a new approach for construction of HCMV mutants. *J Virol*, *73*, 8320-9.

[22] Borst, E. M., Mathys, S., Wagner, M., Muranyi, W. & Messerle, M. (2001). Genetic evidence of an essential role for cytomegalovirus small capsid protein in viral growth. *J Virol*, *75*, 1450-8.

[23] Boyd, M., Safrin, S. & Kern, E. (1993). Penciclovir: A review of its spectrum of activity, selectivity, and cross-resistance pattern. *Antiviral Chemistry & Chemotherapy*, *4*, 3-11.

[24] Breathnach, C. C., Yeargan, M. R., Sheoran, A. S. & Allen, G. P. (2001). The mucosal humoral immune response of the horse to infective challenge and vaccination with equine herpesvirus-1 antigens. *Equine Vet J*, *33*, 651-7.

[25] Browning, G. F., Ficorilli, N. & Studdert, M. J. (1988). Asinine herpesvirus genomes: comparison with those of the equine herpesviruses. Arch *Virol*, *101*, 183-90.

[26] Brune, W., Menard, C., Hobom, U., Odenbreit, S., Messerle, M. & Koszinowski, U. H. (1999). Rapid identification of essential and nonessential herpesvirus genes by direct transposon mutagenesis. *Nat Biotechnol*, *17*, 360-4.

[27] Brune, W., Messerle, M. & Koszinowski, U. H. (2000). Forward with BACs: new tools for herpesvirus genomics. *Trends Genet*, *16*, 254-9.
[28] Brune, W., Hasan, M., Krych, M., Bubic, I., Jonjic, S. & Koszinowski, U. H. (2001). Secreted virus-encoded proteins reflect murine cytomegalovirus productivity in organs. *J Infect Dis*, *184*, 1320-4.
[29] Brune, W., Menard, C., Heesemann, J. & Koszinowski, U. H. (2001) A ribonucleotide reductase homolog of cytomegalovirus and endothelial cell tropism. *Science*, *291*, 303-5.
[30] Bryant, N. A., Davis-Poynter, N., Vanderplasschen, A. & Alcami, A. (2003). Glycoprotein G isoforms from some alphaherpesviruses function as broad-spectrum chemokine binding proteins. *Embo J*, *22*, 833-46.
[31] Burrows, R., Goodridge, D. & Denyer, M. S. (1984). Trials of an inactivated equid herpesvirus 1 vaccine: challenge with a subtype 1 virus. *Vet Rec*, *114*, 369-74.
[32] Chang, W. L. & Barry, P. A. (2003). Cloning of the full-length rhesus cytomegalovirus genome as an infectious and self-excisable bacterial artificial chromosome for analysis of viral pathogenesis. *J Virol*, *77*, 5073-83.
[33] Chung, C. S., Hsiao, J. C., Chang, Y. S. & Chang, W. (1998). A27L protein mediates vaccinia virus interaction with cell surface heparan sulfate. *J Virol*, *72*, 1577-85.
[34] Coen, D. M., Kosz-Vnenchak, M., Jacobson, J. G., Leib, D. A., Bogard, C. L., Schaffer, P. A., Tyler, K. L. & Knipe, D. M. (1989). Thymidine kinase-negative herpes simplex virus mutants establish latency in mouse trigeminal ganglia but do not reactivate. *Proc Natl Acad Sci*, U S A, *86*, 4736-40.
[35] Coen, D. M. & Schaffer, P. A. (2003). Antiherpesvirus drugs: a promising spectrum of new drugs and drug targets. *Nat Rev Drug Discov*, *2*, 278-88.
[36] Colle, C. F., 3rd., Flowers, C. C. & O'Callaghan, D. J. (1992). Open reading frames encoding a protein kinase, homolog of glycoprotein gX of pseudorabies virus, and a novel glycoprotein map within the unique short segment of equine herpesvirus type, 1. *Virology*, *188*, 545-57.
[37] Colle, C. F. & 3rd, O'Callaghan, D. J. (1995) Transcriptional analyses of the unique short segment of EHV-1 strain Kentucky A. *Virus Genes*, *9*, 257-68.
[38] Cornick, J., Martens, J., Martens, R., Crandell, R., McConnell, S. & Kit, S. (1990). Safety and efficacy of a thymidine kinase negative equine herpesvirus-1 vaccine in young horses. *Can J Vet Res*, *54*, 260-6.
[39] Court, D. L., Sawitzke, J. A. & Thomason, L. C. (2002). Genetic engineering using homologous recombination. *Annu Rev Genet*, *36*, 361-88.
[40] Crabb, B. S., Nagesha, H. S. & Studdert, M. J. (1992). Identification of equine herpesvirus 4 glycoprotein G: a type-specific, secreted glycoprotein. *Virology*, *190*, 143-54.
[41] Crabb, B. S. & Studdert, M. J. (1993). Epitopes of glycoprotein G of equine herpesviruses 4 and 1 located near the C termini elicit type-specific antibody responses in the natural host. *J Virol*, *67*, 6332-8.
[42] Crabb, B. S., MacPherson, C. M., Reubel, G. H., Browning, G. F., Studdert, M. J. & Drummer, H. E. (1995a). A type-specific serological test to distinguish antibodies to equine herpesviruses 4 and 1. *Arch Virol*, *140*, 245-58.
[43] Crabb, B. S. & Studdert, M. J. (1995b). Equine herpesviruses 4 (equine rhinopneumonitis virus) and 1 (equine abortion virus). *Adv Virus Res*, *45*, 153-90.

[44] Csellner, H., Walker, C., Wellington, J. E., McLure, L. E., Love, D. N. & Whalley, J. M. (2000). EHV-1 glycoprotein D (EHV-1 gD) is required for virus entry and cell-cell fusion, and an EHV-1 gD deletion mutant induces a protective immune response in mice. *Arch Virol*, *145*, 2371-85.

[45] Cullinane, A. A., Rixon, F. J. & Davison, A. J. (1988). Characterization of the genome of equine herpesvirus 1 subtype 2. *J Gen Virol*, *69*, (Pt 7), 1575-90.

[46] Darlington, R. W. & Randall, C. C. (1963). The nucleic acid content of equine abortion virus. *Virology*, 19, 322-7.

[47] Davis-Poynter, N., Bell, S., Minson, T. & Browne, H. (1994). Analysis of the contributions of herpes simplex virus type 1 membrane proteins to the induction of cell-cell fusion. *J Virol*, *68*, 7586-90.

[48] Davison, A. J., Eberle, R., Ehlers, B., Hayward, G. S., McGeoch, D. J., Minson, A. C., Pellett, P. E., Roizman, B., Studdert, M. J. & Thiry, E. (2009). The order Herpesvirales. *Arch Virol*, *154*, 171-7.

[49] De Clercq, E. (2003). Clinical potential of the acyclic nucleoside phosphonates cidofovir, adefovir, and tenofovir in treatment of DNA virus and retrovirus infections. *Clin Microbiol Rev*, *16*, 569-96.

[50] De Clercq, E. & Holy, A. (2005). Acyclic nucleoside phosphonates: a key class of antiviral drugs. *Nat Rev Drug Discov*, *4*, 928-40.

[51] De Clercq, E., Brancale, A., Hodge, A. V. & Field, H. J. (2006). Antiviral Chemistry & Chemotherapy's current antiviral agents FactFile 2006 (1st edition). *Antivir Chem Chemother, 17*, 113-66.

[52] Delecluse, H. J., Hilsendegen, T., Pich, D., Zeidler, R. & Hammerschmidt, W. (1998) Propagation and recovery of intact, infectious Epstein-Barr virus from prokaryotic to human cells. *Proc Natl Acad Sci*, U S A *95*, 8245-50.

[53] Dewals, B., Boudry, C., Gillet, L., Markine-Goriaynoff, N., de Leval, L., Haig, D. M. & Vanderplasschen, A. (2006). Cloning of the genome of Alcelaphine herpesvirus 1 as an infectious and pathogenic bacterial artificial chromosome. *J Gen Virol*, *87*, 509-17.

[54] Drummer, H. E., Studdert, M. J &, Crabb, B. S. (1998). Equine herpesvirus-4 glycoprotein G is secreted as a disulphide-linked homodimer and is present as two homodimeric species in the virion. *J Gen Virol*, *79* (Pt 5), 1205-13.

[55] Eberle, R., Zhang, M. & Black, D. H. (1993). Gene mapping and sequence analysis of the unique short region of the simian herpesvirus SA 8 genome. *Arch Virol*, *130*, 391-411.

[56] Edington, N., Bridges, C. G. & Patel, J. R. (1986). Endothelial cell infection and thrombosis in paralysis caused by equid herpesvirus-1: equine stroke. *Arch Virol*, *90*, 111-24.

[57] Edington, N., Smyth, B. & Griffiths, L. (1991). The role of endothelial cell infection in the endometrium, placenta and foetus of equid herpesvirus 1 (EHV-1) abortions. *J Comp Pathol*, *104*, 379-87.

[58] Edington, N., Welch, H. M. & Griffiths, L. (1994). The prevalence of latent Equid herpesviruses in the tissues of 40 abattoir horses. *Equine Vet J*, *26*, 140-2.

[59] Eisenberg, R. J., Ponce de Leon, M., Friedman, H. M., Fries, L. F., Frank, M. M., Hastings, J. C. & Cohen, G. H. (1987). Complement component C3b binds directly to purified glycoprotein C of herpes simplex virus types 1 and 2. *Microb Pathog*, *3*, 423-35.

[60] Elion, G. B. (1993). Acyclovir: discovery, mechanism of action, and selectivity. *J Med Virol Suppl, 1*, 2-6.

[61] Ellis, H. M., Yu, D., DiTizio, T. & Court, D. L. (2001). High efficiency mutagenesis, repair, and engineering of chromosomal DNA using single-stranded oligonucleotides. *Proc Natl Acad Sci*, U S A, *98*, 6742-6.

[62] Engelhardt, T. & Keil, G. M. (1996). Identification and characterization of the bovine herpesvirus 5 US4 gene and gene products. *Virology*, *225*, 126-35.

[63] Ficorilli, N., Studdert, M. J. & Crabb, B. S. (1995). The nucleotide sequence of asinine herpesvirus 3 glycoprotein G indicates that the donkey virus is closely related to equine herpesvirus 1. *Arch Virol*, *140*, 1653-62.

[64] Field, H. J. & Wildy, P. (1978). The pathogenicity of thymidine kinase-deficient mutants of herpes simplex virus in mice. *J Hyg*, (Lond) *81*, 267-77.

[65] Field, H. J. & Whitley, R. J. (2005). Antiviral Chemptherapy. In: Mahy, B.W.J., Meulen, V. (Eds.), *Topley & Wilson's Microbiology and Microbial Infections*, 10th edn, Hodder Arnold, London.

[66] Frampton, A. R., Jr., Stolz, D. B., Uchida, H., Goins, W. F., Cohen, J. B. & Glorioso, J. C. (2007). Equine herpesvirus 1 enters cells by two different pathways, and infection requires the activation of the cellular kinase ROCK1. *J Virol*, *81*, 10879-89.

[67] Friedman, H. M., Cohen, G. H., Eisenberg, R. J., Seidel, C. A. & Cines, D. B. (1984). Glycoprotein C of herpes simplex virus 1 acts as a receptor for the C3b complement component on infected cells. *Nature*, *309*, 633-5.

[68] Fries, L. F., Friedman, H. M., Cohen, G. H., Eisenberg, R. J., Hammer, C. H., Frank, M. M. (1986). Glycoprotein C of herpes simplex virus 1 is an inhibitor of the complement cascade. *J Immunol*, *137*, 1636-41.

[69] Fukushi, H., Tomita, T., Taniguchi, A., Ochiai, Y., Kirisawa, R., Matsumura, T., Yanai, T., Masegi, T., Yamaguchi, T. & Hirai, K. (1997). Gazelle herpesvirus 1: a new neurotropic herpesvirus immunologically related to equine herpesvirus 1. *Virology*, 227, 34-44.

[70] Garre, B., van der Meulen, K., Nugent, J., Neyts, J., Croubels, S., De Backer, P. & Nauwynck, H. (2007). In vitro susceptibility of six isolates of equine herpesvirus 1 to acyclovir, ganciclovir, cidofovir, adefovir, PMEDAP and foscarnet. *Vet Microbiol*, *122*, 43-510

[71] Gerber, S. I., Belval, B. J. & Herold, B. C. (1995). Differences in the role of glycoprotein C of HSV-1 and HSV-2 in viral binding may contribute to serotype differences in cell tropism. *Virology*, *214*, 29-39.

[72] Gruenheid, S., Gatzke, L., Meadows, H. & Tufaro, F. (1993). Herpes simplex virus infection and propagation in a mouse L cell mutant lacking heparan sulfate proteoglycans. *J Virol*, *67*, 93-100.

[73] Han, M. G., Kweon, C. H., Mo, I. P. & Kim, S. J. (2002). Pathogenicity and vaccine efficacy of a thymidine kinase gene deleted infectious laryngotracheitis virus expressing the green fluorescent protein gene. *Arch Virol*, *147*, 1017-31.

[74] Hansen, K., Napier, I., Koen, M., Bradford, S., Messerle, M., Bell, E., Seshadri, L., Stokes, H. W., Birch, D. & Whalley, J. M. (2006). In vitro transposon mutagenesis of an equine herpesvirus 1 genome cloned as a bacterial artificial chromosome. *Arch Virol*, *151*, 2389-405.

[75] Harless, W. & Pusterla, N. (2006). Equine Herpesvirus 1 and 4 Respiratory Disease in the Horse. *Clinical Techniques in Equine Practice*, *5*, 197-202.

[76] Harris, S. L., Frank, I., Yee, A., Cohen, G. H., Eisenberg, R. J. & Friedman, H. M. (1990). Glycoprotein C of herpes simplex virus type 1 prevents complement-mediated cell lysis and virus neutralization. *J Infect Dis*, *162*, 331-7.

[77] Heldens, J. G., Hannant, D., Cullinane, A. A., Prendergast, M. J., Mumford, J. A., Nelly, M., Kydd, J. H., Weststrate, M. W. & van den Hoven, R. (2001). Clinical and virological evaluation of the efficacy of an inactivated EHV1 and EHV4 whole virus vaccine (Duvaxyn EHV1,4). Vaccination/challenge experiments in foals and pregnant mares. *Vaccine*, *19*, 4307-17.

[78] Herold, B. C., WuDunn, D., Soltys, N. & Spear, P. G. (1991). Glycoprotein C of herpes simplex virus type 1 plays a principal role in the adsorption of virus to cells and in infectivity. *J Virol*, *65*, 1090-8.

[79] Herold, B. C. & Spear, P. G. (1994). Neomycin inhibits glycoprotein C (gC)-dependent binding of herpes simplex virus type 1 to cells and also inhibits postbinding events in entry. *Virology*, *203*, 166-71.

[80] Herold, B. C., Visalli, R. J., Susmarski, N., Brandt, C. R. & Spear, P. G. (1994). Glycoprotein C-independent binding of herpes simplex virus to cells requires cell surface heparan sulphate and glycoprotein B. *J Gen Virol*, *75*, (Pt 6): 1211-22.

[81] Herring, C. D., Glasner, J. D. & Blattner, F. R (2003). Gene replacement without selection: regulated suppression of amber mutations in Escherichia coli. *Gene*, *311*, 153-63

[82] Hill, F., Benes, V., Thomasova, D., Stewart, A. F., Kafatos, F. C. & Ansorge, W. (2000). BAC trimming: minimizing clone overlaps. *Genomics*, *64*, 111-3.

[83] Hobom, U., Brune, W., Messerle, M., Hahn, G. & Koszinowski, U. H. (2000). Fast screening procedures for random transposon libraries of cloned herpesvirus genomes: mutational analysis of human cytomegalovirus envelope glycoprotein genes. *J Virol*, *74*, 7720-9.

[84] Horsburgh, B. C., Hubinette, M. M., Qiang, D., MacDonald, M. L. & Tufaro, F. (1999a). Allele replacement: an application that permits rapid manipulation of herpes simplex virus type 1 genomes. *Gene Ther*, *6*, 922-30.

[85] Horsburgh, B. C., Hubinette, M. M. & Tufaro, F. (1999b). Genetic manipulation of herpes simplex virus using bacterial artificial chromosomes. *Methods Enzymol*, *306*, 337-52.

[86] Huang, J., Hartley, C. A., Ficorilli, N. P., Crabb, B. S. & Studdert, M. J. (2005). Glycoprotein G deletion mutants of equine herpesvirus 1 (EHV1; equine abortion virus) and EHV4 (equine rhinopneumonitis virus). *Arch Virol*, *150*, 2583-92.

[87] Huang Ja, J. A., Ficorilli, N., Hartley, C. A., Allen, G. P. & Studdert, M. J. (2002). Polymorphism of open reading frame 71 of equine herpesvirus-4 (EHV-4) and EHV-1. *J Gen Virol*, *83*, 525-31.

[88] Huemer, H. P., Larcher, C., Dierich, M. P. & Falke, D. (1992). Factors influencing the interaction of herpes simplex virus glycoprotein C with the third component of complement. *Arch Virol*, *127*, 291-303.

[89] Huemer, H. P., Wang, Y., Garred, P., Koistinen, V. & Oppermann, S. (1993). Herpes simplex virus glycoprotein C: molecular mimicry of complement regulatory proteins by a viral protein. *Immunology*, *79*, 639-47.

[90] Huemer, H. P., Nowotny, N., Crabb, B. S., Meyer, H. & Hubert, P. H. (1995). gp13 (EHV-gC): a complement receptor induced by equine herpesviruses. *Virus Res*, *37*, 113-26.

[91] Hussein, I. T., Menashy, R. V. & Field, H. J. (2008). Penciclovir is a potent inhibitor of feline herpesvirus-1 with susceptibility determined at the level of virus-encoded thymidine kinase. *Antiviral Res*, *78*, 268-74.

[92] Jamsai, D., Orford, M., Nefedov, M., Fucharoen, S., Williamson, R. & Ioannou, P. A. (2003). Targeted modification of a human beta-globin locus BAC clone using GET Recombination and an I-Scei counterselection cassette. *Genomics*, *82*, 68-77.

[93] Jenssen, H., Andersen, J. H., Mantzilas, D. & Gutteberg, T. J. (2004). A wide range of medium-sized, highly cationic, alpha-helical peptides show antiviral activity against herpes simplex virus. *Antiviral Res*, *64*, 119-26.

[94] Johnson, D. C. & Ligas, M. W. (1988). Herpes simplex viruses lacking glycoprotein D are unable to inhibit virus penetration: quantitative evidence for virus-specific cell surface receptors. *J Virol*, *62*, 4605-12.

[95] Kaashoek, M. J., Moerman, A., Madic, J., Rijsewijk, F. A., Quak, J., Gielkens, A. L., van Oirschot, J. T. (1994). A conventionally attenuated glycoprotein E-negative strain of bovine herpesvirus type 1 is an efficacious and safe vaccine. *Vaccine*, *12*, 439-44.

[96] Kanegae, Y., Lee, G., Sato, Y., Tanaka, M., Nakai, M., Sakaki, T., Sugano, S., Saito, I (1995). Efficient gene activation in mammalian cells by using recombinant adenovirus expressing site-specific Cre recombinase. *Nucleic Acids Res*, *23*, 3816-21.

[97] Karger, A. & Mettenleiter, T. C. (1993). Glycoproteins gIII and gp50 play dominant roles in the biphasic attachment of pseudorabies virus. *Virology*, *194*, 654-64.

[98] Karger, A., Saalmuller, A., Tufaro, F., Banfield, B. W. & Mettenleiter, T. C. (1995). Cell surface proteoglycans are not essential for infection by pseudorabies virus. *J Virol*, *69*, 3482-9.

[99] Kavanagh, D. G., Gold, M. C., Wagner, M., Koszinowski, U. H., Hill, A. B. (2001) The multiple immune-evasion genes of murine cytomegalovirus are not redundant: m4 and m152 inhibit antigen presentation in a complementary and cooperative fashion. *J Exp Med*, *194*, 967-78.

[100] Keil, G. M., Engelhardt, T., Karger, A. & Enz, M. (1996). Bovine herpesvirus 1 U(s) open reading frame 4 encodes a glycoproteoglycan. *J Virol*, *70*, 3032-8.

[101] Keller, K. M., Brauer, P. R. & Keller, J. M. (1989). Modulation of cell surface heparan sulfate structure by growth of cells in the presence of chlorate. *Biochemistry*, *28*, 8100-7.

[102] Kempkes, B., Pich, D., Zeidler, R., Sugden, B. & Hammerschmidt, W. (1995). Immortalization of human B lymphocytes by a plasmid containing 71 kilobase pairs of Epstein-Barr virus DNA. *J Virol*, *69*, 231-8.

[103] Kit, S., Kit, M. & Pirtle, E. C. (1985a). Attenuated properties of thymidine kinase-negative deletion mutant of pseudorabies virus. *Am J Vet Res*, *46*, 1359-67.

[104] Kit, S., Qavi, H., Gaines, J. D., Billingsley, P. & McConnell, S. (1985b). Thymidine kinase-negative bovine herpesvirus type 1 mutant is stable and highly attenuated in calves. *Arch Virol*, *86*, 63-83.

[105] Klupp, B. G., Karger, A. & Mettenleiter, T. C. (1997). Bovine herpesvirus 1 glycoprotein B does not productively interact with cell surface heparan sulfate in a pseudorabies virion background. *J Virol*, *71*, 4838-41.

[106] Klupp, B. G., Nixdorf, R. & Mettenleiter, T. C. (2000). Pseudorabies virus glycoprotein M inhibits membrane fusion. *J Virol*, *74*, 6760-8.

[107] Kydd, J. H., Smith, K. C., Hannant, D., Livesay, G. J. & Mumford, J. A. (1994). Distribution of equid herpesvirus-1 (EHV-1) in respiratory tract associated lymphoid tissue: implications for cellular immunity. *Equine Vet J*, *26*, 470-3.

[108] Liang, X. P., Babiuk, L. A., van Drunen Littel-van den Hurk, S., Fitzpatrick, D. R. & Zamb, T. J. (1991). Bovine herpesvirus 1 attachment to permissive cells is mediated by its major glycoproteins gI., gIII, and gIV. *J Virol*, *65*, 1124-32.

[109] Ligas, M. W. & Johnson, D. C. (1988). A herpes simplex virus mutant in which glycoprotein D sequences are replaced by beta-galactosidase sequences binds to but is unable to penetrate into cells. *J Virol*, *62*, 1486-94.

[110] Lindahl, U., Kusche, M., Lidholt, K. & Oscarsson, L. G. (1989). Biosynthesis of heparin and heparan sulfate. *Ann N Y Acad Sci*, *556*, 36-50.

[111] Lubinski, J. M., Wang, L., Soulika, A. M., Burger, R., Wetsel, R. A., Colten, H., Cohen, G. H., Eisenberg, R. J., Lambris, J. D. & Friedman, H. M. (1998). Herpes simplex virus type 1 glycoprotein gC mediates immune evasion in vivo. *J Virol*, *72*, 8257-63.

[112] Lycke, E., Johansson, M., Svennerholm, B. & Lindahl, U. (1991). Binding of herpes simplex virus to cellular heparan sulphate, an initial step in the adsorption process. *J Gen Virol*, *72*, (Pt 5): 1131-7.

[113] Maeda, K., Yokoyama, N., Fujita, K., Maejima, M. & Mikami, T. (1997). Heparin-binding activity of feline herpesvirus type 1 glycoproteins. *Virus Res*, *52*, 169-76.

[114] Maeda, K., Yasumoto, S., Tsuruda, A., Andoh, K., Kai, K., Otoi, T. & Matsumura, T (2007). Establishment of a novel equine cell line for isolation and propagation of equine herpesviruses. *J Vet Med Sci*, *69*, 989-91.

[115] Mahony, T. J., McCarthy, F. M., Gravel, J. L., West, L. & Young, P. L. (2002). Construction and manipulation of an infectious clone of the bovine herpesvirus 1 genome maintained as a bacterial artificial chromosome. *J Virol*, *76*, 6660-8.

[116] Matsumura, T., Sugiura, T., Imagawa, H., Fukunaga, Y. & Kamada, M. (1992). Epizootiological aspects of type 1 and type 4 equine herpesvirus infections among horse populations. *J Vet Med Sci*, *54*, 207-11.

[117] Matsumura, T., O'Callaghan, D. J., Kondo, T. & Kamada, M. (1996). Lack of virulence of the murine fibroblast adapted strain, Kentucky A (KyA), of equine herpesvirus type 1 (EHV-1) in young horses. *Vet Microbiol*, *48*, 353-65.

[118] Matsumura, T., Kondo, T., Sugita, S., Damiani, A. M., O'Callaghan, D. J. & Imagawa, H. (1998). An equine herpesvirus type 1 recombinant with a deletion in the gE and gI genes is avirulent in young horses. *Virology*, *242*, 68-79.

[119] McGeoch, D. J., Moss, H. W., McNab, D. & Frame, M. C. (1987). DNA sequence and genetic content of the HindIII l region in the short unique component of the herpes simplex virus type 2 genome: identification of the gene encoding glycoprotein G, and evolutionary comparisons. *J Gen Virol*, *68*, (Pt 1): 19-38.

[120] McGregor, A. & Schleiss, M. R. (2001). Molecular cloning of the guinea pig cytomegalovirus (GPCMV) genome as an infectious bacterial artificial chromosome (BAC) in Escherichia coli. *Mol Genet Metab*, *72*, 15-26.

[121] McGregor, A. & Schleiss, M. R. (2001). Recent advances in herpesvirus genetics using bacterial artificial chromosomes. *Mol Genet Metab*, *72*, 8-14.

[122] McGregor, S., Easterday, B. C., Kaplan, A. S. & Ben-Porat, T. (1985). Vaccination of swine with thymidine kinase-deficient mutants of pseudorabies virus. *Am J Vet Res*, *46*, 1494-7.

[123] McNearney, T. A., Odell, C., Holers, V. M., Spear, P. G. & Atkinson, J. P. (1987). Herpes simplex virus glycoproteins gC-1 and gC-2 bind to the third component of complement and provide protection against complement-mediated neutralization of viral infectivity. *J Exp Med*, *166*, 1525-35.

[124] Melancon, J. M., Luna, R. E., Foster, T. P. & Kousoulas, K. G. (2005). Herpes simplex virus type 1 gK is required for gB-mediated virus-induced cell fusion, while neither gB and gK nor gB and UL20p function redundantly in virion de-envelopment. *J Virol*, *79*, 299-313.

[125] Messerle, M., Crnkovic, I., Hammerschmidt, W., Ziegler, H. & Koszinowski, U. H. (1997) Cloning and mutagenesis of a herpesvirus genome as an infectious bacterial artificial chromosome. *Proc Natl Acad Sci*, U S A *94*, 14759-63.

[126] Mettenleiter, T. C., Zsak, L., Zuckermann, F., Sugg, N., Kern, H. & Ben-Porat, T. (1990). Interaction of glycoprotein gIII with a cellular heparinlike substance mediates adsorption of pseudorabies virus. *J Virol*, *64*, 278-86.

[127] Mettenleiter, T. C., Klupp, B. G., Weiland, F. & Visser, N. (1994). Characterization of a quadruple glycoprotein-deleted pseudorabies virus mutant for use as a biologically safe live virus vaccine. *J Gen Virol*, *75*, (Pt 7): 1723-33.

[128] Mittal, S. K. & Field, H. J. (1989). Analysis of the bovine herpesvirus type 1 thymidine kinase (TK) gene from wild-type virus and TK-deficient mutants. *J Gen Virol*, *70*, (Pt 4), 901-18.

[129] Montgomery, R. I., Warner, M. S., Lum, B. J. & Spear, P. G. (1996). Herpes simplex virus-1 entry into cells mediated by a novel member of the TNF/NGF *receptor family*., *Cell*, *87*, 427-36.

[130] Morfin, F. & Thouvenot, D. (2003). Herpes simplex virus resistance to antiviral drugs. *J Clin Virol*, *26*, 29-37.

[131] Muyrers, J. P., Zhang, Y., Testa, G. & Stewart, A. F. (1999). Rapid modification of bacterial artificial chromosomes by ET-recombination. *Nucleic Acids Res*, *27*, 1555-7.

[132] Muyrers, J. P., Zhang, Y. & Stewart, A. F. (2000). ET-cloning: think recombination first. *Genet Eng*, (N Y) *22*, 77-98.

[133] Muyrers, J. P., Zhang, Y. & Stewart, A. F. (2001). Techniques: Recombinogenic engineering--new options for cloning and manipulating DNA. *Trends Biochem Sci*, *26*, 325-31.

[134] Nagaike, K., Mori, Y., Gomi, Y., Yoshii, H., Takahashi, M., Wagner, M., Koszinowski, U. & Yamanishi, K. (2004). Cloning of the varicella-zoster virus genome as an infectious bacterial artificial chromosome in Escherichia coli. *Vaccine*, *22*, 4069-74.

[135] Nagy, A. (2000). Cre recombinase: the universal reagent for genome tailoring. *Genesis*, *26*, 99-109.

[136] Narayanan, K., Williamson, R., Zhang, Y., Stewart, A. F. & Ioannou, P. A. (1999). Efficient and precise engineering of a 200 kb beta-globin human/bacterial artificial chromosome in E. coli DH10B using an inducible homologous recombination system. *Gene Ther*, *6*, 442-7.

[137] Neubauer, A., Braun, B., Brandmuller, C., Kaaden, O. R. & Osterrieder, N. (1997). Analysis of the contributions of the equine herpesvirus 1 glycoprotein gB homolog to virus entry and direct cell-to-cell spread. *Virology*, *227*, 281-94.

[138] Neubauer, A. & Osterrieder, N. (2004). Equine herpesvirus type 1 (EHV-1) glycoprotein K is required for efficient cell-to-cell spread and virus egress. *Virology*, *329*, 18-32.

[139] O'Keefe, J. S., Murray, A., Wilks, C. R. & Moriarty, K. M. (1991). Amplification and differentiation of the DNA of an abortigenic (type 1) and a respiratory (type 4) strain of equine herpesvirus by the polymerase chain reaction. *Res Vet Sci*, *50*, 349-51.

[140] Okazaki, K., Matsuzaki, T., Sugahara, Y., Okada, J., Hasebe, M., Iwamura, Y., Ohnishi, M., Kanno, T., Shimizu, M. & Honda, E. et al. (1991). BHV-1 adsorption is mediated by the interaction of glycoprotein gIII with heparinlike moiety on the cell surface., *Virology, 181*, 666-70.

[141] Osterrieder, N. (1999). Construction and characterization of an equine herpesvirus 1 glycoprotein C negative mutant. *Virus Res*, *59*, 165-77.

[142] Parry, C., Bell, S., Minson, T. & Browne, H. (2005). Herpes simplex virus type 1 glycoprotein H binds to alphavbeta3 integrins. *J Gen Virol*, *86*, 7-10.

[143] Patel, J. R., Foldi, J., Bateman, H., Williams, J., Didlick, S. & Stark, R. (2003). Equid herpesvirus (EHV-1) live vaccine strain C147: efficacy against respiratory diseases following EHV types 1 and 4 challenges. *Vet Microbiol*, *92*, 1-17.

[144] Patel, J. R. & Heldens, J. (2005). Equine herpesviruses 1 (EHV-1) and 4 (EHV-4)--epidemiology, disease and immunoprophylaxis: a brief review. *Vet J*, *170*, 14-23.

[145] Pusterla, N., Leutenegger, C. M., Wilson, W. D., Watson, J. L., Ferraro, G. L. & Madigan, J. E. (2005). Equine herpesvirus-4 kinetics in peripheral blood leukocytes and nasopharyngeal secretions in foals using quantitative real-time TaqMan PCR. *J Vet Diagn Invest*, *17*, 578-81

[146] Qiu, J., Handa, A., Kirby, M. & Brown, K. E. (2000). The interaction of heparin sulfate and adeno-associated virus 2. *Virology*, *269*, 137-47.

[147] Ramsay, M. (1994). Yeast artificial chromosome cloning. *Mol Biotechnol*, *1*, 181-201.

[148] Rea, T. J., Timmins, J. G., Long, G. W. & Post, L. E. (1985). Mapping and sequence of the gene for the pseudorabies virus glycoprotein which accumulates in the medium of infected cells. *J Virol*, *54*, 21-9.

[149] Reed, S. M. & Toribio, R. E. (2004). Equine herpesvirus 1 and 4. Vet *Clin North Am Equine Pract*, *20*, 631-42.

[150] Roizman, B. & Sears, A. E. (1987). An inquiry into the mechanisms of herpes simplex virus latency. *Annu Rev Microbiol*, *41*, 543-71.

[151] Roizman, B. (1996). Herpesviridae. In: B. N., Field, D. M., Knipe, P. M., Howley, R. M., Channock, J. L., Melnick, T. P., Monath, B. Roizman, & S. E. Straus, (Eds.)., *Virology*, third edn. Lippincott-Raven, Philadelphia, NY.

[152] Roizmann, B., Desrosiers, R. C., Fleckenstein, B., Lopez, C., Minson, A. C. & Studdert, M. J. (1992). The family Herpesviridae: an update. The Herpesvirus Study Group of the International Committee on Taxonomy of Viruses. *Arch Virol 123*, 425-49.

[153] Rollinson, E. A. & White, G. (1983). Relative activities of acyclovir and BW759 against Aujeszky's disease and equine rhinopneumonitis viruses. *Antimicrob Agents Chemother, 24*, 221-6.

[154] Rudolph, J., O'Callaghan, D. J. & Osterrieder, N. (2002). Cloning of the genomes of equine herpesvirus type 1 (EHV-1) strains KyA and racL11 as bacterial artificial chromosomes (BAC). *J Vet Med B Infect Dis Vet Public Health*, *49*, 31-6.

[155] Rudolph, J. & Osterrieder, N. (2002). Equine herpesvirus type 1 devoid of gM and gp2 is severely impaired in virus egress but not direct cell-to-cell spread. *Virology*, *293*, 356-67.

[156] Sambrook, J. & Russell, D. W. (2001). *Molecular Cloning: a Laboratory Manual.* Cold Spring Harbor, NY: Cold Spring Harbor Laboratory.

[157] Sanchez, V., Clark, C. L., Yen, J. Y., Dwarakanath, R. & Spector, D. H. (2002). Viable human cytomegalovirus recombinant virus with an internal deletion of the IE2 86 gene affects late stages of viral replication. *J Virol*, *76*, 2973-89.

[158] Satoh, T., Arii, J., Suenaga, T., Wang, J., Kogure, A., Uehori, J., Arase, N., Shiratori, I., Tanaka, S., Kawaguchi, Y., Spear, P. G., Lanier, L. L. & Arase, H. (2008). PILRalpha is a herpes simplex virus-1 entry coreceptor that associates with glycoprotein *B. Cell*, *132*, 935-44.

[159] Schalkwyk, L. C., Francis, F. & Lehrach, H. (1995). Techniques in mammalian genome mapping. *Curr Opin Biotechnol*, *6*, 37-43.

[160] Schreurs, C., Mettenleiter, T. C., Zuckermann, F., Sugg, N. & Ben-Porat, T. (1988). Glycoprotein gIII of pseudorabies virus is multifunctional. *J Virol*, *62*, 2251-7.

[161] Schumacher, D., Tischer, B. K., Fuchs, W. & Osterrieder, N. (2000). Reconstitution of Marek's disease virus serotype 1 (MDV-1) from DNA cloned as a bacterial artificial chromosome and characterization of a glycoprotein B-negative MDV-1 mutant. *J Virol*, *74*, 11088-98.

[162] Sergerie, Y. & Boivin, G. (2006). Thymidine Kinase Mutations Conferring Acyclovir Resistance in Herpes Simplex Type 1 Recombinant Viruses. *Antimicrob Agents Chemother.*, 3889-3892.

[163] Sharma, P. C., Cullinane, A. A., Onions, D. E. & Nicolson, L (1992). Diagnosis of equid herpesviruses -1 and -4 by polymerase chain reaction. *Equine Vet J, 24*, 20-5.

[164] She, K. (2003). So you want to work with giants: the BAC vector. *BioTeach Journal*, *1*, 69-74.

[165] Shieh, M. T., WuDunn, D., Montgomery, R. I., Esko, J. D. & Spear, P. G. (1992). Cell surface receptors for herpes simplex virus are heparan sulfate proteoglycans. *J Cell Biol, 116*, 1273-81.

[166] Shizuya, H., Birren, B., Kim, U. J., Mancino, V., Slepak, T., Tachiiri, Y. & Simon, M. (1992). Cloning and stable maintenance of 300-kilobase-pair fragments of human DNA in Escherichia coli using an F-factor-based vector. *Proc Natl Acad Sci*, U S A *89*, 8794-7.

[167] Shizuya, H. & Kouros-Mehr, H. (2001). The development and applications of the bacterial artificial chromosome cloning system. *Keio J Med*, *50*, 26-30.

[168] Shukla, D., Liu, J., Blaiklock, P., Shworak, N. W., Bai, X., Esko, J. D., Cohen, G. H., Eisenberg, R. J., Rosenberg, R. D. & Spear, P. G. (1999). A novel role for 3-O-sulfated heparan sulfate in herpes simplex virus 1 entry. *Cell*, *99*, 13-22.

[169] Slater, J. D., Gibson, J. S. & Field, H. J. (1993). Pathogenicity of a thymidine kinase-deficient mutant of equine herpesvirus 1 in mice and specific pathogen-free foals. *J Gen Virol*, 74, (Pt 5)., 819-28.

[170] Slater, J. D., Lunn, D. P., Horohov, D. W., Antczak, D. F., Babiuk, L., Breathnach, C., Chang, Y. W., Davis-Poynter, N., Edington, N., Ellis, S., Foote, C., Goehring, L., Kohn, C. W., Kydd, J., Matsumura, T., Minke, J., Morley, P., Mumford, J., Neubauer, T., O'Callaghan, D., Osterrieder, K., Reed, S., Smith, K., Townsend, H., Meulen, Kvd., Whalley, M. & Wilson, W. D. (2006). Report of the equine herpesvirus-1 Havermeyer Workshop, San Gimignano, Tuscany, June 2004. *Veterinary Immunology and Immunopathology*, *111*, 3-13.

[171] Slomka, M. J., Harrington, L., Arnold, C., Norcott, J. P. & Brown, D. W. (1995). Complete nucleotide sequence of the herpesvirus simiae glycoprotein G gene and its expression as an immunogenic fusion protein in bacteria. *J Gen Virol*, *76*, (Pt 9).: 2161-8.

[172] Smith, D. J., Iqbal, J., Purewal, A., Hamblin, A. S. & Edington, N. (1998). In vitro reactivation of latent equid herpesvirus-1 from CD5+/CD8+ leukocytes indirectly by IL-2 or chorionic gonadotrophin. *J Gen Virol*, *79*, (Pt 12).: 2997-3004.

[173] Smith, G. A. & Enquist, L. W. (1999). Construction and transposon mutagenesis in Escherichia coli of a full-length infectious clone of pseudorabies virus, an alphaherpesvirus. *J Virol*, *73*, 6405-14.

[174] Smith, G. A. & Enquist, L. W. (2000). A self-recombining bacterial artificial chromosome and its application for analysis of herpesvirus pathogenesis. *Proc Natl Acad Sci*, U S A, *97*, 4873-8

[175] Smith, G. R. (1988). Homologous recombination in procaryotes. *Microbiol Rev*, *52*, 1-28

[176] Smith, G. R. (2001). Homologous recombination near and far from DNA breaks: alternative roles and contrasting views. *Annu Rev Genet*, *35*, 243-74.

[177] Spatz, S. J., Rota, P. A. & Maes, R. K. (1994). Identification of the feline herpesvirus type 1 (FHV-1) genes encoding glycoproteins G, D, I and E: expression of FHV-1 glycoprotein D in vaccinia and raccoon poxviruses. *J Gen Virol*, *75*, (Pt 6): 1235-44.

[178] Spear, P. (1985). Glycoproteins specified by herpes simplex virus. In *The Herpesviruses* (B. Roizman, ed.), *Vol. 3*, 315-356. Plenum, New York.

[179] Spear, P. G. (1993). Entry of alphaherpesviruses into cells. *Seminars in Virology*, *4*, 167-180.

[180] Spear, P. G. & Longnecker, R. (2003). Herpesvirus entry: an update. *J Virol*, *77*, 10179-85.

[181] Spear, P. G. (2004). Herpes simplex virus: receptors and ligands for cell entry. *Cell Microbiol*, *6*, 401-10.

[182] Stanberry, L. R., Kit, S. & Myers, M. G. (1985). Thymidine kinase-deficient herpes simplex virus type 2 genital infection in guinea pigs. *J Virol*, *55*, 322-8.

[183] Stokes, A., Allen, G. P., Pullen, L. A. & Murray, P. K. (1989). A hamster model of equine herpesvirus type 1 (EHV-1) infection; passive protection by monoclonal antibodies to EHV-1 glycoproteins 13, 14 and 17/18. *J Gen Virol*, *70*, (Pt 5): 1173-83.

[184] Studdert, M. J. (1974). Comparative aspects of equine herpesviruses. *Cornell Vet*, *64*, 94-122

[185] Suter, M., Lew, A. M., Grob, P., Adema, G. J., Ackermann, M., Shortman, K. & Fraefel, C. (1999). BAC-VAC, a novel generation of (DNA) vaccines: A bacterial artificial chromosome (BAC) containing a replication-competent, packaging-defective

virus genome induces protective immunity against herpes simplex virus 1. *Proc Natl Acad Sci*, U S A, *96*, 12697-702.

[186] Tanaka, M., Kagawa, H., Yamanashi, Y., Sata, T. & Kawaguchi, Y. (2003). Construction of an excisable bacterial artificial chromosome containing a full-length infectious clone of herpes simplex virus type 1: viruses reconstituted from the clone exhibit wild-type properties in vitro and in vivo. *J Virol*, *77*, 1382-91.

[187] Tearle, J. P., Smith, K. C., Platt, A. J., Hannant, D., Davis-Poynter, N. J. & Mumford, J. A. (2003). In vitro characterisation of high and low virulence isolates of equine herpesvirus-1 and -4. *Res Vet Sci*, *75*, 83-6.

[188] Telford, E. A., Watson, M. S., McBride, K. & Davison, A. J. (1992). The DNA sequence of equine herpesvirus-1. *Virology*, *189*, 304-16.

[189] Telford, E.A ., Watson, M. S., Perry, J., Cullinane, A. A. & Davison, A. J. (1998). The DNA sequence of equine herpesvirus-4. *J Gen Virol*, *79*, 1197-203.

[190] Tischer, B. K., von Einem, J., Kaufer, B. & Osterrieder, N. (2006). Two-step red-mediated recombination for versatile high-efficiency markerless DNA manipulation in Escherichia coli. *Biotechniques*, *40*, 191-7.

[191] Tischer, B. K., Kaufer, B. B., Sommer, M., Wussow, F., Arvin, A. M. & Osterrieder, N. (2007). A self-excisable infectious bacterial artificial chromosome clone of varicella-zoster virus allows analysis of the essential tegument protein encoded by ORF9. *J Virol*, *81*, 13200-8.

[192] Trapp, S., Osterrieder, N., Keil, G. M. & Beer, M. (2003). Mutagenesis of a bovine herpesvirus type 1 genome cloned as an infectious bacterial artificial chromosome: analysis of glycoprotein E and G double deletion mutants. *J Gen Virol*, *84*, 301-6.

[193] Van de Walle, G. R., May, M. L., Sukhumavasi, W., von Einem, J. & Osterrieder, N. (2007). Herpesvirus chemokine-binding glycoprotein G (gG). efficiently inhibits neutrophil chemotaxis in vitro and in vivo. *J Immunol*, *179*, 4161-9.

[194] Van de Walle, G. R., Peters, S. T., VanderVen, B. C., O'Callaghan, D. J. & Osterrieder, N. (2008). Equine herpesvirus 1 entry via endocytosis is facilitated by alphaV integrins and an RSD motif in glycoprotein D. *J Virol*, *82*, 11859-68.

[195] Van de Walle, G. R., Sakamoto, K. & Osterrieder, N. (2008). CCL3 and viral chemokine-binding protein gg modulate pulmonary inflammation and virus replication during equine herpesvirus 1 infection. *J Virol*, *82*, 1714-22.

[196] van der Meulen, K., Garre, B., Croubels, S. & Nauwynck, H. (2006). In vitro comparison of antiviral drugs against feline herpesvirus 1. *BMC Vet Res*, *2*, 13.

[197] van Maanen, C. (2002). Equine herpesvirus 1 and 4 infections: an update. *Vet Q*, *24*, 58-78.

[198] Voziyanov, Y., Pathania, S. & Jayaram, M. (1999). A general model for site-specific recombination by the integrase family recombinases. *Nucleic Acids Res*, *27*, 930-41.

[199] Wagner, M., Jonjic, S., Koszinowski, U. H. & Messerle, M. (1999). Systematic excision of vector sequences from the BAC-cloned herpesvirus genome during virus reconstitution. *J Virol*, *73*, 7056-60.

[200] Wagner, M., Michel, D., Schaarschmidt, P., Vaida, B., Jonjic, S., Messerle, M., Mertens, T. & Koszinowski, U. (2000). Comparison between human cytomegalovirus pUL97 and murine cytomegalovirus (MCMV) pM97 expressed by MCMV and vaccinia virus: pM97 does not confer ganciclovir sensitivity. *J Virol*, *74*, 10729-36.

[201] Wagner, M., Ruzsics, Z. & Koszinowski, U. H. (2002). Herpesvirus genetics has come of age. *Trends Microbiol*, *10*, 318-24.

[202] Warming, S., Costantino, N., Court, D. L., Jenkins, N. A. & Copeland, N. G. (2005). Simple and highly efficient BAC recombineering using galK selection. *Nucleic Acids Res*, *33*, e36.

[203] Whalley, J. M., Ruitenberg, K. M., Sullivan, K., Seshadri, L., Hansen, K., Birch, D., Gilkerson, J. R. & Wellington, J. E. (2007). Host cell tropism of equine herpesviruses: glycoprotein D of EHV-1 enables EHV-4 to infect a non-permissive cell line. *Arch Virol*, *152*, 717-25.

[204] Wilkins, P. (2004). Acyclovir in the treatment of EHV-1 myeloencephalopathy. 22nd annual meeting of the American college of veterinary internal medicine proceeding, 170-172.

[205] Wilson, W. D. (1997). Equine herpesvirus 1 myeloencephalopathy. Vet Clin *North Am Equine Pract*, *13*, 53-72.

[206] Wong, Q. N., Ng, V. C., Lin, M. C., Kung, H. F., Chan, D. & Huang, J. D. (2005). Efficient and seamless DNA recombineering using a thymidylate synthase A selection system in Escherichia coli. *Nucleic Acids Res, 33*, e59.

[207] WuDunn, D. & Spear, P. G. (1989). Initial interaction of herpes simplex virus with cells is binding to heparan sulfate. *J Virol*, *63*, 52-8.

[208] Yang, X. W., Model, P. & Heintz, N. (1997). Homologous recombination based modification in Escherichia coli and germline transmission in transgenic mice of a bacterial artificial chromosome. *Nat Biotechnol*, *15*, 859-65.

[209] Yang, Y. & Sharan, S. K. (2003). A simple two-step, 'hit and fix' method to generate subtle mutations in BACs using short denatured PCR fragments. *Nucleic Acids Res*, *31*, e80.

[210] Yu, D., Ellis, H. M., Lee, E. C., Jenkins, N. A., Copeland, N. G. & Court, D. L. (2000). An efficient recombination system for chromosome engineering in Escherichia coli. *Proc Natl Acad Sci,* U S A *97*, 5978-83.

[211] Yu, D., Smith, G. A., Enquist, L. W. & Shenk, T. (2002). Construction of a self-excisable bacterial artificial chromosome containing the human cytomegalovirus genome and mutagenesis of the diploid TRL/IRL13 gene. *J Virol*, *76*, 2316-28.

[212] Zhang, Y., Buchholz, F., Muyrers, J. P. & Stewart, A. F. (1998). A new logic for DNA engineering using recombination in Escherichia coli. *Nat Genet*, *20*, 123-8.

[213] Zhang, Z., Rowe, J., Wang, W., Sommer, M., Arvin, A., Moffat, J. & Zhu, H. (2007). Genetic analysis of varicella-zoster virus ORF0 to ORF4 by use of a novel luciferase bacterial artificial chromosome system. *J Virol*, *81*, 9024-33.

[214] Zhao, Y., Petherbridge, L., Smith, L. P., Baigent, S. & Nair, V. (2008). Self-excision of the BAC sequences from the recombinant Marek's disease virus genome increases replication and pathogenicity. *Virol J*, *5*, 19.

[215] Zhou, F. C., Zhang, Y. J., Deng, J. H., Wang, X. P., Pan, H. Y., Hettler, E. & Gao, S. J. (2002). Efficient infection by a recombinant Kaposi's sarcoma-associated herpesvirus cloned in a bacterial artificial chromosome: application for genetic analysis. *J Virol*, *76*, 6185-96.

[216] Zhu, Z., Gershon, M. D., Ambron, R., Gabel, C. & Gershon, A. A. (1995). Infection of cells by varicella zoster virus: inhibition of viral entry by mannose 6-phosphate and heparin. *Proc Natl Acad Sci*, U S A, *92*, 3546-50.

In: Horses: Biology, Domestication, and Human Interactions
Editor: James E. Leffhalm
ISBN: 978-1-61761-303-6

Chapter 2

A REVIEW OF THE SELECTIVE BREEDING IN HORSES

***Javad Bakhtiari*[*1] *and Giti Heshmat*[2]**

[1] Animal Genetics and Breeding; Graduated of University of Tehran, Iran
[2]Animal Science Department, Islamic Azad University, Mahabad, Iran

ABSTRACT

In ancient world the horses were kept because they provided food as well as materials for clothing and shelter for our ancestors. In the modern world the most part of the horses are kept as a hobby and they are participated in racing, jumping, showing and etc. The horse's physical structure is the result of evolution, controlled initially by environment as well as selective breeding enforced by human. Evidence of selective breeding of horses for both riding and draught purposes exists from ancient Elam in 2800 BC. Development of the racing horses has been founded based on the fact that a horse's ability to run a given distance in a relatively short time is partly determined by genetic components. In selective breeding we need to define a breeding goal and some criteria to measure the performance of an individual animal. In race horses the criteria used to estimate the racing performance are time traits, handicap weight, performance rate, rank at finish, earning prize and the like. In this chapter we want to describe the history of the selection in horses, briefly, and to identify the criteria which have been used for measuring of the performance of the racing horses and might be used to maximize improvement of racing performance as well as to introduce some other criteria that can be used in selecting of the parents of racing horses, also, investigate the advantages and disadvantages of these criteria.

[*] Corresponding author: Tel.: +98 914 320 1352; E-mail address: bakhtiarijavad@yahoo.com

INTRODUCTION

From the time that man domesticated the horses, they have been used for different range of functions. In the modern world the most part of the horses are kept as a hobby and they are participated in racing, jumping, showing and etc. The differences found among breeds of the horses are the result of evolution, controlled initially by environment as well as artificial selection enforced by human. However, the evidence indicates that whenever human began to interfere with the process of nature, it was the man that increasingly influenced the subsequent development of the horse.

Race horse breeds have been created mainly by using selective breeding in a relatively long period of time. Generally horse breeders have spent more time and money to obtain high performances from their horses. By using proper selection criteria, profitability of horse breeding can be improved. In this chapter we have focused on the studies that have been carried out on Thoroughbreds, Trotters and Arab horses. Thoroughbred horse racing is the most diffused form of horse racing throughout the world [Thiruvenkadan et al., 2009 a]. The Thoroughbred is the fastest of the world's breeds (Edwards, 2000) and has the best racing time over the so called classic distance of 1.5 miles or 2400 m. The trotter ranks worldwide second to the Thoroughbred in popularity as a racehorse [Thiruvenkadan et al., 2009 b]. Arab horse races, also, become to be popular more and more in some countries. In contrast with Thoroughbreds the Arab horses begin to race at the age of three and their average race distance is longer. In this chapter we want to describe the history of the selection in horses, briefly, and to identify the criteria which have been used for measuring the performance of the racing horses and might be used to maximize improvement of racing performance as well as to introduce some other criteria that can be used in selecting of the parents of racing horses, also, investigate the advantages and disadvantages of these criteria.

THE HISTORY OF DOMESTICATION AND SELECTIVE BREEDING OF THE HORSES

Evidence shows that the horse being domesticated in Eurasia about 5-6,000 years ago [Edwards, 2000; Outram et al., 2009]. In ancient world the horses were kept because they provided food as well as materials for clothing and shelter for our ancestors. Their flesh provided food; their hides were used to make tents and clothes; and the manure could be dried to make fires. Mares supplied milk, which could be fermented into kummis, the fiery brew of the steppes [Edwards, 2000].

From the time of domestication the horses have been used for different purposes such as riding, communication and agricultural progress, transporting in war or peace, sport and so on. As the methods of agriculture improved, horses could be hand feed. This combined with selective breeding, produced bigger, stronger and faster horses suited to the particular requirements of the day. For the most part, those requirements were related with warfare and transport, as well as, of course, for the purposes of sport in the great circuses of the classical civilization of Greece and Rome [Edwards, 2000].

Evidence of selective breeding of the horses for both riding and draught purposes exists from ancient Elam in 2800 BC [Amschler, 1935]. By 400 BC, selection standards, remarkably similar to those used for selection of modern riding and racing horses, had been defined for cavalry horses [the hittite warrior Kikuli's treatise 3500 BP, Simeon of athena 424 BC, Xenophon, 369 BC; Evans et al., 1977].

Selective breeding has an important role in development of horse breeds which have been selected over relatively long periods of time. By gelding of male horses the use of selection in horse breeding has been possible and mating of the horses from best stocks increased the quality and elevated the breed's purposed characters. Arab horse population, the purest and probably the oldest of all breeds, provide the best example of the breed which has been bred and carefully selected over thousands of years for developing a unique beauty, a character that would not have been possible without artificial selection planned by human [Edwards, 2000]. Development of the racing horses has been founded based on the facts that racing performance is inheritable and a horse's ability to run a given distance in a relatively short time is partly determined by genetic components. Development of Thoroughbreds, which evolved in England as a result of the crossing between Oriental sires and a native stock of running horses [Edwards, 2000; Cunningham, 2005] and have been selected for speed since Henry VIII of England, and American Standard breed, which was developed from successfully raced Thoroughbred seed stock and was officially designated as a breed in 1892, confirm these beliefs [Tolley et al., 1985].

SELECTION CRITERIA

In selective breeding we need to define a breeding goal and some criteria to measure the performance of an individual animal. Performance of the race horses can be measured by some different criteria. However, there is little agreement in selecting the criteria for measuring the performance of the race horses. In fact, each country defines its selection goal thorough the establishment and management of racing programs and uses it as a tool for performance testing.

Running ability in Thoroughbreds was defined as a composite trait, resulted from speed, stamina and competitive temperament [stone, 1935]. Steele (1944) studied the performance of the Thoroughbreds and noted that racing capacity was often expressed in terms of an individual's ability to win stake races.

In picking the selection criterion some factors such as the heritability of the trait, facility of measuring and using of the records, suitability of the records and trait, breeding goal and selected trait convergence, and the like should be considered. Regardless the problems of selecting the criteria, performance of the race horses can be measured by some traits such as: time traits, handicap weights, performance rates, rank at finish, earning prize and so on. These traits reflect an individual horse's ability to respond to a particular set of conditions.

Conformation traits, also, can be used in selection of racing horses because they are associated to the desirable performance traits of the horses. In race horses it is possible to use early measured conformation traits for prediction of their future racing performances and, therefore, their earlier selection [Smith et al., 2006]. Bourgelat (1754) used a quantitative approach to conformation traits and studied linear measurements of the body segments of the baroque horses. Despite of the importance of conformation traits, unfortunately, the information about these traits in race horses is too little and it is not comprehensive.

Time Traits

A natural choice to measure a performance trait in race horses is to measure its ability to run fast and speed is a traditional measure used as a selection criterion. Racing time can be defined as a time that was elapsed over a given distance. In each races, racing time is the only direct measure of the speed and the only suitable quantitative measure that can be used for evaluating of the racing performance of the horses without any transforming. However, times recorded during a race are affected by the specific conditions of that race. In other word the environmental factors such as age, sex, rider, trainer, origin of the horse, type of the track, condition of the track, the horses participated in that race (class of the race), post position number that a horse start its competition from it, distance of the race, handicap weight, total carried weight, the amount of the prize, differences between the prize of the horses with different rank and etc. influence the racing time of the horses belong to different breeds or within one breed. Furthermore, in some races, racing time is recorded only for winner horses of each race and not for all the horses participated in that race [Mota et al., 2005].

Regardless of those problems, however, racing time is the most frequently reported measure of racing performance. The estimates of the heritability of racing time in different studies vary widely according to the correction factors and ranging from near to zero to 0.78 in different breeds and populations (see Table 1). Most of the reported heritabilities in various studies are lower than 0.20 and in some cases are even very low. Chico (1994) reported very low and near to zero (0.01) heritability for speed of the Spanish Thoroughbreds. Bakhtiari and Kashan (2009) reported the estimates of 0.09, to 0.13 for racing time of Iranian Thoroughbred horses in different distances. In contrast, Ekiz and Kocak (2005) reported relatively high heritabilities of 0.35 and 0.28 for racing time of Arab horses respectively in dirt track and based on entire dataset in turkey. In the same study the estimate of heritability of racing time in turf track was 0.18.

Since recorded racing time during a race can be influenced by some factors associated with that race, several authors have attempted to adjust their records for various sources of variation.

In a time trial, a horse is forced to attain the highest rate of its speed that is possible for a given distance. The obtained records are not necessarily correlated with the results of a true competition affected by the additional factors may interact with the ability of the horse. However, time trial data may indicate an upper level of achievement [Tolley et al., 1985].

Best time achieved during a racing year or the lifetime best time may also be used as a selection criterion of the racing horses [Tolley et al., 1985]. Neisser (1976) reported the heritability of 0.13 for best time of 1533 progeny of the 41 Thoroughbred sires. In that study average time was not significantly affected by track condition, track or month, and was significantly correlated with best time (0.72). Ekiz and Kocak (2005) also reported the heritability of 0.23, 0.36 and 0.28 for best racing time in turf, dirt tracks and based on entire data set respectively.

The reported results of some authors in Thoroughbreds [Oki et al., 1995; Oki and Sasaki, 1995; Williamson and Beilharz, 1998; Mota et al., 2005; Ekiz and Kocak, 2007; Bakhtiari and Kashan, 2009] indicated that the estimates of the heritability and repeatability of racing time tended to decrease with increasing racing distance. With increase of racing distance more environmental factors (such as the effect of rider's tactics) influence the racing performance of the horses [Oki and Sasaki, 1995].

Table 1. heritability (h^2) and repeatability (r^2) estimates for time related traits

Author(s) and date	Trait	Breed	h2 (±SE)	r2 (±SE)
Artz (1961)	Average corrected time	Thoroughbred	0.19 and 0.24	
Dusek (1963)	Racing time	Thoroughbred	0.25 and 0.45	
Bormann (1964)	Average corrected time for:	Thoroughbred		
	2- year olds		0.17	
	3- year olds		0.09	
	4- year olds		0.17	
Dusek (1965)	Racing time	Thoroughbred	0.20 and 0.25	
Bormann (1966)	Corrected time for:	Thoroughbred		
	2- year olds		0.06	0.19
	3- year olds		0.14	0.34
Watanabe (1969)	Average corrected time	Thoroughbred	0.12	
Pern (1970)	Time (reduction/Km)	Thoroughbred	0.53	
Pern (1971)	Speed	Thoroughbred	0.05 - 0.08	
Pern (1973)	Speed (corrected)	Thoroughbred	0.10 - 0.78	
Neisser (1976)	Best racing time	Thoroughbred	0.13	
Neisser and Schwark (1979)	Racing time	Thoroughbred	0.10	
Chico (1994)	Speed	Thoroughbred	0.01	0.01
Oki et al. (1995)	Racing time at different distances in:	Thoroughbred		
	Dirt track		0.09 - 0.22	0.51 - 0.67
	Turf track		0.08 - 0.25	0.43 - 0.70
Taveira et al. (2004)	Racing time	Thoroughbred	0.08	
Ekiz et al. (2005) [27]	Racing time for:	Thoroughbred		
	Entire dataset		0.32 (±0.022)	0.36
	Dirt track		0.40 (±0.029)	0.47
	Turf track		0.23 (±0.025)	0.29
	Best racing time for			
	Entire dataset		0.47 (±0.044)	0.50
	Dirt track		0.49 (±0.045)	0.53
	Turf track		0.29 (±0.035)	0.37
Mota et al. (2005)	Racing time at different distances	Thoroughbred	0.04 - 0.29	0.15 - 0.63
Mota (2006)	Racing time at different distances	Thoroughbred	0.10 - 0.32	
Ekiz and Kocak (2007)	Racing time at different distances	Thoroughbred	0.18 - 0.35	0.29 - 0.40
Bakhtiari and Kashan (2009)	Racing time at different distances	Thoroughbred	0.09 - 0.13	0.17 - 0.26
Ekiz et al. (2005) [26]	Racing time at different distances	Arabian	0.18 - 0.30	0.30 - 0.46
Ekiz and Kocak (2005)	Racing time for:	Arabian		
Kalmykov (1973)	Racing time	Trotter	0.34	

Table 1. (Continued)

Author(s) and date	Trait	Breed	h2 (±SE)	r2 (±SE)
Katona (1979)	timing	Trotter		
	Best time / year		0.14 - 0.31	
	Average racing time / year		0.21 - 0.49	
	Entire dataset		0.28 (±0.031)	0.42
	Dirt track		0.35 (±0.038)	0.50
	Turf track		0.18 (±0.031)	0.35
	Best racing time for			
	Entire dataset		0.28 (±0.042)	0.43
	Dirt track		0.36 (±0.050)	0.52
	Turf track		0.23 (±0.045)	0.45
Bugislaus et al. (2005)	Racing time/Km	Trotter	0.19 and 0.23	0.47 and 0.53

In longer distances, the riders of the horses in front, can slow the speed of the horse and contribute more environmental differences into the racing time of the same horses in their different distances and races. This is the contribution of a non-permanent environment which causes the lower heritability values if not adjusted by a race effect. One can also note that, there must be a relatively short distance (such as 1000 m) which a horse can run fast merely as its biological morphology and physiology make it possible [Mota et al., 2005].

Also, all horses can compete in short races. Therefore short races contain not only true sprinters but also stayers primarily racing to gain fitness for their future races. Furthermore, young horses start their racing in sprint events and compete at short distances before the decision to race them at long-distance categories or to retire them from the track. Poor performers never progress to longer events. Thus the longer distances contain a selected population and the genetic variation among horses and hence the heritability estimates should be lower than existing in the sprint races where sprinters and stayers compete [Sobczynska, 2006]. These results show that selection based on racing time in short distances can be more successful than long distances and can be result in more genetic improvement. However, in contrast, Williamson and Beilharz (1996) reported higher heritability for longer distances in Thoroughbred.

Genetic correlations between racing time at different distances, reported in various studies, are moderate and with increasing the differences between racing distances, the estimates of genetic correlations decreased [Oki et al., 1997; Mota, 2006; Bakhtiari and Kashan, 2009]. This indicates that the racing time at different distances are not exactly the same trait and with the selection of the horses based on one distance we cannot improve the performance of the next generation in all distances.

Bormann (1966), reported low correlations between corrected mean time and the handicap weight for 2 (0.25) and 3-year olds (0.5) Thoroughbreds that clearly show the relative independency of timing and success in races. This demonstrates that mean speed and handicap weight are not exactly the same trait and measuring the same ability.

Handicap Weights

In Thoroughbred horses, they are handicapped by assigning increasingly heavier weights to the horses as their supposed racing merit increases. Stepwise increases in weight carried have the average effect of slowing the speed of an individual over a given distance [Laughlin, 1934 a,b]. Despite of the fact that handicap weights are subjective measurements, some studies have been carried out and used handicaps as an indicator of level of performance to avoid the problem of quantifying relative performance of different horses in many races during a year [Tolley et al., 1985].

Estimates of heritability of handicap weights reported in different studies range from 0.10 to 0.93 and frequently are between 0.20 and 0.50 (Table 2). Some authors reported the evidence of bias such as effects of a common environment on parents and offspring [More O'Ferrall, and Cunningham, 1974] and effects of assortative matings [Field and Cunningham, 1976]. However, when the handicap weights are generalized to all individuals of a population, they give a reliable estimation of the horses to be used in selection. It was suggested that handicap weights should be preferred to timings, as their heritability seems substantially higher [Bormann, 1966] if it is not overestimated [Langlois, 1980].

Table 2. heritability (h^2) estimates of handicap weights in Thoroughbreds

Author(s) and date	h^2
Dusek (1963)	0.25 and 0.45
Dusek (1965)	0.19 and 0.25
Bormann (1966)	0.33 and 0.51
Schwark and Neisser (1971)	0.60
More O'Ferrall and Cunningham (1974)	0.34 - 0.74
Hecker (1975)	0.23 - 0.40
Field and Cunningham (1976)	0.23 - 0.93
Bodo (1976)	0.10 - 0.51

Performance Rate

Performance rate can be defined as the number of lengths that a horse is expected to finish in front of, or behind an average horse in an average race in a given year or season. Gillespie (1971) suggested that since performance rates are based on the number of lengths ahead, or behind an average horse in an average race, it is possible to make comparisons between horses using performance rates. For example a horse with a rate of 15.0 would, on the average, be expected to finish five lengths ahead of one with a rate of 10.0. He proposed that it is better to separate the horses with different age groups. However in this way comparing the performance of the horses with different ages is not possible because they are running separately.

In generating the performance rates, each horse receives a score for every start during the season, and its performance rate is the average of all such scores. The score of an individual horse for a single performance is calculated by adding to the rating of the race the average number of lengths by which the losing horses were defeated, then subtracting the distance by which the individual horse was defeated. In the case of the winner, the subtraction amounts to zero lengths. The rating of the race is the average of the performance rates of the entire season or year of all the competing horses [Robertson, 1971].

Since race rating depend upon the ratings of the competing horses, and the horses are rated in terms of the races, a number of adjustments should be done: going back to adjust a race rating as the performance rates of the horses which contested it change during the season, then coming forward to adjust the performance rates in terms of the adjusted race rates, and so forth. This iteration process is continued, by computer, until both sets of ratings stop changing [Robertson, 1971].

In order to convert the arbitrary distances “nose”, “head” and “neck” into numerical lengths the following definitions have been made: nose = 0.05 lengths, head = 0.15 lengths, and neck = 0.30 lengths [Gillespie, 1971].

In calculating this criterion, factors such as length of the race, weight carried, post position, rider, track condition, and so on are not considered. However, performance rates are objective measures based on actual performance of the horses rather than handicap weights that are subjective and appraised by race judges.

Heritability of performance rate estimated in various studies ranges from 0.36 to 0.94 (Table 3). Watanabe (1974), studied the performance rate of the Japanese Thoroughbreds and reported the heritability of 0.64 for this trait. In other study Kieffer (1975), studied performance rates of 3-year olds Thoroughbred horses and suggested that environmental correlations within sire progenies can be result in overestimated heritability for it.

Table 3. heritability (h^2) estimates of performance rate in Thoroughbreds

Author(s) and date	h^2 (±SE)
Foye et al. (1972)	0.36 - 0.68
Watanabe (1974)	0.64 (±0.23)
Kieffer (1975)	0.39 – 1.45

Earning Prize

Earning prize is another criterion for measuring the performance of the race horses. Estes (1948), used the Average Earnings Index (AEI) for sire selection, which compared active sires with the average of the breed and ranked the sires according to performance of their progeny based on the amount of the prize earned per starter per year for each $1000 available. He concluded that racing merit can be equated with AEI. Ever since then, most of the researchers have used the earnings as a criterion for measuring the performance of the race horses.

The reasons of using earnings as a performance indicator described as: all individuals have an equal opportunity, but not ability to win; earnings indicated what an individual was able to do in actual competition; and, therefore, level of earnings implied level of performance [Estes, 1952].

In the study of performance of Thoroughbred horses, Estes (1952) used the AEI as a measure of racing ability and reported a significant correlation between dam and progeny of 100 selected mares. However, when the data was adjusted for the effects of performance of the dam, temperament of dam, sex of progeny and all two-way interactions, only the effect of the sex of progeny was significant on performance of the offspring.

Clark (1977) separated earnings of fillies from those of the colts and geldings and used the Standard Starts Index (SSI). He, also, ranked 2-year olds horses separately from the older.

Estimates of the heritability and repeatability of earning prize reported in studies range from 0.08 to 0.60 and from 0.23 to 0.45 respectively. The significancy of age, sex, year of the race, origin of the horse and breeder on earnings have reported by some authors [Belhajyahia et al., 2003; Sobczynska and Lukaszewicz, 2003; Sobczynska and Lukaszewicz, 2004; Ekiz and Kocak, 2005; Ekiz et al., 2005 (27)]. Sobczynska and Lukaszewicz (2004), also, noted the importance of the corrections of earnings for yearly inflation and the specific attributes of the races such as prize size and class of the races.

There are some difficulties in using the earning prize as a measure of the racing performance. Distribution of the records of earning prize is not normal and it may affect the accuracy of the estimates. Therefore, some researchers normalized earnings by a Log transformation. Langlois (1975) studied the earnings of 3-year old Thoroughbreds. He adjusted the earnings and average earnings per start by Estes' method [Estes, 1948] and normalized them by Log transformation. Estimations of heritability in that study ranged from 0.02 to 0.56, depended on the used criteria and methods of estimation (Table 4).

Later in a study of genetic parameters of racing merit in Poland Thoroughbred horses, Sobczynska and Lukaszewicz (2004) suggested that a model including the results of individual races is better than a model including only simple averaged records, because the former introduces the additional fixed effect of the race. Moreover, simple averaging is not sufficient to account for a complex specificity of a particular race and it may result in particularly bias evaluation of horses which have had few progeny or have raced seldom.

Another problem is the fact that prize money, in each races, is given only to the placed (first three to five) horses; so the rest of the horses remain non-earned. The first choice for non-placed non-earned horses is deleting the records of them from data set or considering them as miss data. Sobczynska and Lukaszewicz (2004) asserted that choice of "0" to non earning horses artificially decreases the variance of the records and including the horses with "0" earning, also, leads to deformation of the trait distribution. Moreover, they concluded that because considering the "0" does not follow the differentiation of endowments in different level races, it is not a good choice. In this way, a non placed horse in a difficult race with high endowment is equated with a horse that not earned any prize in an easier race; and non earned horses in a prestigious race are underestimated.

In a study of racing performance of Thoroughbreds in Spain, to solve these problems, Chico (1994) used artificial earnings for non earned horses which were calculated by multiplying the earnings by 0.5 for each increasing rank. In the other study, Svobodova et al. (2005) used the same method for calculating the non-real earnings. After normalization by Log transformation, they equalized the non-real earnings by assigning the mean of the Log of non-real earnings. However, it seems that equalization of the non-real earnings may result in a skew distribution of the data.

Table 4. heritability (h^2) and repeatability (r^2) estimates of earnings

Author(s) and date	Trait	Breed	h^2 (±SE)	r^2 (±SE)
Pirri and Steele (1952) [63]	Log of total earnings	Thoroughbred	0.60	
Pirri and Steele (1952) [64]	Log of total earnings	Thoroughbred	0.14	
Watanabe et al. (1966)	Earnings	Thoroughbred	0.45	
Foye et al. (1972)	Yearly average earning/start	Thoroughbred	0.30	
Langlois (1975)	Earnings at 3 years	Thoroughbred	0.04 – 0.07	
	Earnings per start at 3 years		0.02 – 0.06	
	Log of earnings at 3 years		0.24 – 0.44	
	Log of earnings per start at 3 years		0.40 – 0.56	
Galizzi-Vecciotti and Pazzaglia (1976)	Log of Estes' average earnings/start	Thoroughbred	0.12 - 0.29	
Chico (1994)	Log of earnings	Thoroughbred	0.08 and 0.09	0.23 and 0.25
Sobczynska and Lukaszewicz (2004)	Log of earnings	Thoroughbred	0.10 (± 0.12)	0.23 (± 0.23)
Ekiz et al. (2005) [27]	Annual earnings	Thoroughbred	0.19 (± 0.039)	0.32
	Earnings/start		0.29 (± 0.046)	0.45
	Log of earnings		0.19 (± 0.036)	0.31
	Log of earnings/start		0.34 (± 0.042)	0.41
Svobodova et al. (2005)	Log of earnings/race for:	Thoroughbred		
	all starting horses		0.14 (± 0.010)	0.31 (± 0.014)
	restricted to placed horses		0.19 (± 0.008)	0.40 (± 0.016)
	Log of annual earnings for:			
	two year olds		0.15 (± 0.029)	
	three year olds		0.34 (± 0.033)	
	2+3 year olds		0.32 (± 0.026)	
Belhajyahia et al (2003)	Log of earnings	Arabian	0.09 (± 0.011)	0.25 (± 0.006)
Sobczynska and Lukaszewicz (2003)	Log of earnings	Arabian	0.14	0.29
Ekiz and Kocak (2005)	Annual earnings	Arabian	0.14 (± 0.027)	0.30
	Earnings/start		0.17 (± 0.017)	0.38
	Log of earnings		0.15 (± 0.031)	0.34
	Log of earnings/start		0.17 (± 0.039)	0.32
Katona (1979)	Total earnings	Trotter	0.11 (± 0.02)	
	Earnings/start		0.25 (± 0.04)	
Bugislaus et al. (2005)	Log of earnings/race	Trotter	0.08 and 0.09	0.28 and 0.29

In most of the races, decreasing the earnings between placed horses is not regular according to increasing the rank of the horses. Furthermore, the earnings awarded to the placed horses don't cover different levels of the races efficiently. These problems were discussed by Belhajyahia et al. (2003) who studied the breeding value of Arab horses in Tunisia. They concluded that estimated genetic parameters resulted from normalized ranking value are more accurate than those of Log of earnings. Because of high genetic and phenotypic correlations (0.97 and 0.78 respectively) between Log of earnings and normalized ranking and higher heritability and repeatability estimates, they suggested choosing normalized ranking value as a measurement criterion of performance.

Ranking Value

In the recent studies, placing or rank at finish has been used more and more as a measure of racing performance of race horses. Similar to the earnings ranking is not a normal variable. Chico (1994), normalized ranking records of Spanish thoroughbred horses by replacing them with a normal score introduced by Pearson and Hartley (1972) which was the expected value of the rank out of n individuals on an underlying standardized normal variable. Using two models, she reported the estimates of 0.07 and 0.10 for heritability of normalized ranking value. Some other authors used the square root of the rank records for normalizing the values [Sobczynska and Lukaszewicz, 2003; Sobczynska and Lukaszewicz, 2004; Bugislaus et al., 2005].

Estimates of heritability and repeatability of ranking, reported in different studies, vary from 0.07 to 0.25 and from 0.13 to 0.47 respectively (Table 5). In various studies, the effects of the race, age, sex, origin of the horse and weight carried on ranking values were significant [Chico, 1994; Belhajyahia et al., 2003; Ekiz and Kocak, 2005; Ekiz et al., 2005 (27)].

Sobczynska (2006) studied genetic correlations between racing performance of Thoroughbreds and Arab horses at different racing distances in Poland. Genetic correlations between rankings at two different distances, estimated in that study, were between 0.54 and 0.98 and between 0.95 and 0.99 for Thoroughbred and Arab horses respectively. He concluded that rankings at different distance races are the same traits for Arabs but not exactly the same for Thoroughbreds. The similar result was reported by Bakhtiari and Kashan (2009) for Iranian Thoroughbred horses.

Other Criteria for Measuring the Performance of Race Horses

In few studies the authors used some other criteria for measuring the performance of the race horses and selecting the parents of the next generation. In this part we, briefly, introduce some of these criteria.

In the study of heritability of racing merit in Arab horses, Sobczynska and Lukaszewicz (2003) used the **number of annual starts** (square root) of the horses as a measure of racing merit. It was noted that since fitness and soundness help the horses achieving frequent starts, the number of annual starts, therefore, can be considered as an indicator of animal's toughness. Heritability of this trait was reported as 0.12 in that study.

Table 5. Estimates of heritability (h^2) and repeatability (r^2) of ranking

Author(s) and date	Trait	Breed	h^2 (±SE)	r^2 (±SE)
Neisser and Schwark (1979)	Placing	Thoroughbred	0.10	
Chico (1994)	Normalized ranking value	Thoroughbred	0.07 and 0.10	0.21 - 0.26
Sobczynska and Lukaszewicz (2004)	Ranking (square root)	Thoroughbred	0.15 (± 0.18)	0.33 (± 0.34)
Ekiz et al. (2005) [27]	Ranking for:	Thoroughbred		
	Entire dataset		0.13 (±0.016)	0.23
	Dirt track		0.13 (±0.019)	0.23
	Turf track		0.15 (±0.020)	0.29
Svobodova et al. (2005)	Normalized ranking value for:	Thoroughbred		
	all starting horses		0.16 (± 0.011)	0.35 (± 0.014)
	placed horses		0.17 (± 0.008)	0.32 (± 0.016)
Sobczynska (2006)	Ranking (square root) for:			
	Thoroughbreds	Thoroughbred	0.04 - 0.16	0.16 - 0.39
	Arab horses	Arabian	0.02 - 0.08	0.20 - 0.38
Bakhtiari and Kashan (2009)	Normalized ranking value in different distances	Thoroughbred	0.11 - 0.17	0.20 - 0.32
Belhajyahia et al (2003)	Normalized ranking value	Arabian	0.12 (± 0.013)	0.35 (± 0.007)
Sobczynska and Lukaszewicz (2003)	Ranking (square root)	Arabian	0.25	0.47
Ekiz and Kocak (2005)	Ranking for:	Arabian		
	Entire dataset		0.07 (±0.011)	0.13
	Dirt track		0.07 (±0.012)	0.13
	Turf track		0.10 (±0.019)	0.17
Bugislaus et al. (2005)	Ranking (square root)	Trotter	0.05 and 0.07	0.20 and 0.24

Precocity is another criterion that some researchers have used it in race horses. In some countries such as Finland, Germany and Sweden, precocity has been selected as a breeding goal [Mota and Oliveira, 2000].

The possibility to make early performance and the relation between early performance and later performances are two aspects of the precocity [Ricard, 1998]. Low starting age and ability of good performance at a young age improve the profitability of horse breeding and keeping [Mota and Oliveira, 2000]. Moreover, the horses starting their races at an early age are healthier than those beginning at older ages [Darenius et al., 1983], as well as, they can start more often and may have better racing results [Saastamoinen and Ojala, 1991].

Early performance: Saastamoinen and Nylander (1996 a,b) studied age at the first start of trotters. Estimated heritabilities for age at first qualifying start, passed qualifying start and first race of trotters were 0.04 to 0.16. Some researchers noted that the age of the horses at first start may be affected by some factors such as slow development and, therefore, maturing age, intentions of owners or trainers, injuries, and lack of talent [Bendroth, 1981; Klemetsdal et al. 1985; Saastamoinen, 1991].

Relation with later performances: in a study Mota and Oliveira (2000), estimated the genetic and phenotypic correlations between age of the horses at the first race (F) and race track performance traits, related with their later performances, (number of victories (NV), number of placements in top 5 (N5), percentage of victories (PV), percentage of placements in top 5 (P5) and number of starts (NS)) in Brazilian Thoroughbred race horses. According to the results of the study, it was concluded that the horses with the high genetic merit for NV and particularly PV, were those with a low genetic potential for F. the highest phenotypic correlations were reported between F:N5 (-0.13) and F:NS (-0.15) and the other correlations were very low and closed to zero.

Longevity: in race horses longevity can be defined as the length of racing life of an individual horse. Because of the great amount of invested time and money, longevity in race horses is a trait of considerable economic importance [Wallin et al., 2001]. Number of factors such as conformation of the horses, sex, age, trainer, type of competition and track conditions may affect the longevity of race horses.

In Thoroughbreds, injury is the main reason of omitting the horses [Rossdale et al., 1985; Robinson and Gordon, 1988] and it was reported that poor conformation can lead to a horse injury [Stashak, 2002]. Fatality rate of the race horses was studied by McKee (1995) in the UK. He reported that the fatality rates in national hunt flat, hurdle and steeplechase races were 0.47, 0.49 and 0.70% respectively. In the other study, comparing with the horses racing on the flat races, Bailey et al. (1998) reported that horses in hurdle races were about four times as likely to suffer musculoskeletal broken, while horses in steeplechases had eight times more chance of suffering an injury. Burns et al. (2006), estimated the heritability of longevity in Thoroughbreds and reported the heritability of 0.12 based on complete dataset analysis.

Conformation Traits

Conformation traits include body measurements and angels, leg stances, hoof quality and movements (their length, elasticity and regularity) [Saastamoinen and Barrey, 2000]. These traits can be divided to objective or measured (i.e. body measurements and angels) and subjective or scored (i.e. hoof quality, leg stances and movements) traits. The former are measured objectively using the equipments such as photography, video techniques or other equipments. Scored traits are evaluated by judging and giving points for their correctness. In this chapter we have reviewed the measured traits.

In race horses, conformation of the body, mainly, is the result of artificial selection planned by human, as well as, natural selection. Conformation of the body varies widely among different breeds and even between different subpopulations and lines of the same breed [Saastamoinen and Barrey, 2000].

Conformation traits are interested for horse breeders because they are associated to the desirable characterization and performance traits of the animals [Barbosa, 1993]. In racing horses conformation traits are correlated with their racing performance. For example, withers height of mature horses is positively correlated with racing performance [Dolvik and Klemetsdal, 1999] and stride length [Galisteo et al., 1998]. In Thoroughbred horses, also, yearling measurements can be used to predict mature dimensions with reasonable accuracy [Anderson and Mcllwraith, 2004; Staniar et al., 2004], therefore it is logical to assume that conformation measurements of yearling Thoroughbreds may have predictive value for their future racing performances [Smith et al., 2006]. Moreover, well balanced animals achieve better market prices. Therefore, these traits are important and can be used in evaluation and selection of racing horses.

Table 6. Estimates of heritability (h^2) for conformation traits

Author(s) and date	Trait	Breed	h^2 (±SE)
Bakhtiari and Heshmat		Thoroughbred	
(2009)	Head length		0.39 (±0.043)
	Length of back line		0.41 (±0.028)
	Length of shoulder		0.24 (±0.036)
	Heart girth		0.49 (±0.047)
	Chest width		0.22 (±0.024)
	Withers height		0.38 (±0.018)
	Front cannon height		0.27 (±0.031)
	Circumference of front cannon		0.30 (±0.052)
	Hind cannon height		0.29 (±0.023)
	Circumference of hind cannon		0.34 (±0.018)
Gharahveysi et al.		Arabian	
(2008)	Head length		0.37 (±0.094)
	Neck length		0.05 (±0.008)
	Withers height		0.57 (±0.092)
	Mid body height		0.43 (±0.093)
	Croup height		0.61 (±0.087)
	Body length		0.27 (±0.087)
	Back cannon circumference		0.19 (±0.055)
	Fore cannon circumference		0.06 (±0.009)
	Fore cannon bone length		0.13 (±0.099)
	Heart girth circumference		0.26 (±0.060)
	Chest width		0.13 (±0.074)
	Chest depth		0.54 (±0.095)
	Umbilical circumference		0.24 (±0.065)
Saastamoinen et al.		Trotter	
(1998)	Withers height		0.78 (±0.02)
	Croup height		0.77 (±0.03)
	Body length		0.64 (±0.02)
	Chest width		0.50 (±0.02)
	Cannon bone circumference		0.65 (±0.02)

The studies about genetic parameters of conformation traits in race horses and their correlations with racing performance of the horses are too little. Some traits such as withers height, heart girth, chest width and circumference of cannon are most studied. On the other hand, there is very little information for some other traits such as head length, shoulder length and cannon height in race horses.

Reported heritability estimates for conformation traits in most of the breeds are moderate to high (Table 6). This indicates that mass selection can be used for these traits and with doing selection, a suitable response will be found. Most of the genetic correlations between these traits, reported in various studies, are also moderate to high [Costa et al. 1998; Molina et al, 1999; Smith et al., 2006; Bakhtiari and Heshmat, 2009], meaning that selection for one of these traits should result in an increase in the other traits of interest.

ACKNOWLEDGMENT

The authors are grateful to Dr. Bertrand Langlois for reviewing this manuscript and Frank Columbus for giving us this opportunity.

REFERENCES

Amschler, W. (1935). The oldest pedigree chart. *Journal of heredity*, *26*, 233.

Anderson, T. M. & Mcllwraith, C. W. (2004). Longitudinal development of equine conformation from weanling to age 3 years in the Thoroughbred. *Equine Vet. J.*, *36*, 563-570.

Artz, W. (1961). A contribution on the evaluation of performance tests in Thoroughbred breeding with special reference to the racing performance of individual stallion progeny groups. *Giess. Schriftenr. Tierz. Haustiergenet.*, *2*, 1-62.

Bailey, C. J., Reid, S. W., Hodgson, D. R. & Bourke, J. M. et al. (1998). Flat, hurdle and steeple racing: risk factors for musculoskeletal injury. *Equine vet. J.*, *30*, 498-503. (Cited in Burns et al., 2006).

Bakhtiari, J. & Heshmat, G. (2009). Estimation of genetic parameters of conformation traits in Iranian Thoroughbred horses. *Livestock Science*, *123,* 116-120.

Bakhtiari, J. & Kashan, N. E. J. (2009). Estimation of genetic parameters of racing performance in Iranian Thoroughbred horses. *Livestock Science*, *120,* 151-157.

Barbosa, C. G. (1993). Disertacao de Mestrado, Universidade federal de Minas Gerais, Brazil. (Cited in Costa et al., 1998).

Belhajyahia, T., Blouin, C., Langlois, B. & Harzalla, H. (2003). Breeding evaluation of Arab horses from their racing results in Tunisia by a BLUP with an animal model approach. *Anim. Res.*, *52,* 481-488.

Bendroth, M. (1981). A survey of reasons for some being non-starters as 2, 3 and 4 years olds. *Paper, 32nd Ann. Meet. Of Eur. Assoc. for Anim. Prod.*

Bodo, I. (1976). Critical considerations on variable estimates of the degree of inheritance in a race horse population. *Fed. Eur. Zootech. (Comm. Chev.) 27th Congr., Zurich.*

Bormann, P. (1964). Die Anwendung biomathematischer Methoden bei der Auswertung der Rennleistung von Vollblutpferden. *Giessener Schriftenreihe, Tierzucht und Haustiergenetik* 7.

Bormann, P. (1966). Ein Vergleich zwischen Generalausgleichsgewicht und Zeitmessung als Selektionsmasstab in der Vollblutzucht. *Zuchtungskunde, 38,* 301-310.

Bourgelat (1754). A *New System of Horsemanship from the French Monsieur* Bourgelat Brenger, A. (Ed.), Henry Woodfall, London. (Cited in Saastamoinen, & Barrey, 2000).

Bugislaus, A. E., Roehe, R. & Kalm, E. (2005). Comparison of two different statistical models considering individual races or racetracks for evaluation of German trotters. *Livestock Production Science, 92,* 69-76.

Burns, E. M., Enns, R. M. & Garrick, D. J. (2006). The effect of simulated censored data on estimates of heritability of longevity in the Thoroughbred racing industry. *Genet. Mol. Res., 5(1)*, 7-15.

Chico, M. D. (1994). Genetic analysis of Thoroughbred racing performance in Spain. *Ann Zootech., 43,* 393-397.

Clark, S. (1977). S.S.I. *The Blood Horse,* 1612-1617. Cited in Langlois (1980).

Cunningham, E. P. (2005). Molecular methods in equine genetic diversity. *EAAP Publication, 116,* 15-24.

Darenius, A., Bendroth, M., Arnason, T. & Phillipsson, J. (1983). Tavlingsstatistik som underlag for genetiska studer av travarens hallbarhet. *Svensk Veterinartidning, 35(3)*, 49-51. (Cited in Mota and Oliveira 2000).

Dolvik, N. I. & Klemetsdal, G. (1999). Conformation traits of Norwegian cold-blooded Trotters: heritability and the relationship with performance. *Acta. Agric. Scand, Sect A, Animal Sci., 49,* 156-162.

Dusek, J. (1963). Bemerkungen zur Beurteilung der Leistungsfahigkeit von Pferden. (Observations on the evaluation of performance in horses). *Schriftenr. Max-Plank Inst. Tierz. Tierernaehr.*, Spec., 257-284.

Dusek, J. (1965). Prispevek ke studio dedicnosti nekterych vlastnosti koni. (The heritability of some characters in the horses.) *Zivocisna Vyroba., 6,* 449-456.

Edwards, E. H. (2000). *The ultimate horse book.* London, UK: Dorling Kindersley Limited.

Ekiz, B. & Kocak, O. (2005). Phenotypic and genetic parameter estimates for racing traits of Arabian horses in Turkey. *J. Anim. Breed. Genet., 122,* 349-356.

Ekiz, B. & Kocak, O. (2007). Estimates of genetic parameters for racing times of Thoroughbred horses. *Turk. J. Vet. Anim. Sci., 31(1)*, 1-5.

Ekiz, B., Kocak, O. & Demir, H. (2005). Estimates of genetic parameters for racing performances of Arabian horses. *Turk. J. Vet. Anim. Sci., 29,* 543-549.

Ekiz, B., Kocak, O. & Yilmaz, A. (2005). Phenotypic and genetic parameter estimates for racing traits of Thoroughbred horse in Turkey. *Arch. Tierz., Dummerstorf., 48(2),* 121-129.

Estes, B. W. (1952). A study of the relationship between temperament of Thoroughbred broodmares and performance of offspring. *Journal of Genetic Psychology, 81,* 273.

Estes, J. A. (1948). Statistics on prominent sires, adjusted for changing dollar. *The Blood Horse, 52,* 470.

Evans, B. W., Borton, A., Hintz, F.W. & Van Vleck, L. D. (1977). *The horse.* San Francisco. W.H. Freeman and Company.

Field, J. K. & Cunningham, E. P. (1976). A further study of the inheritance of racing performance in Thoroughbred horses. *Journal of Heredity*, *67,* 247-248.

Foye, D. B., Dickey, H. C. & Sniffen, C. J. (1972). Heritability of racing performance and a selection index for breeding potential in the Thoroughbred horse. *Journal of Animal Science*, *35,* 1141-1145.

Galisteo, A. M., Cano, M. R., Morales, J. L., Vivo, J. & Miro, F. (1998). The influence of speed and height at the withers on the kinematics of sound horses at the hand-led trot. *Vet. Res. Commun.*, *22,* 415-423.

Galizzi-Vecciotti, A. G. & Pazzaglia, G. (1976). Ereditabilita dell'attitudine alla corsa del puro sangue Inglese studiata su corse piane italiane. (Heritability of racing performance in Thoroughbreds studied in Italian flat races.) *Atti delle Societa Italiana delle Scienze Veterinarie.*, *30,* 490-492.

Gharahveysi, S., Kashan, N. E. J., Gerami, A. & Torshizi, R. V. (2008). Estimation of genetic parameters on conformation traits of the Iranian Arab Horses Population. *Pakistan Journal of Biological Sciences*, *11(2)*, 280-284.

Gillespie, R. H. (1971). A new way to evaluate race horses, performance rates. *The Thoroughbred Record*, *17,* 961-977.

Hecker, W. (1975). The heritability of speediness. *Allattenyesztes.*, *24,* 117-121.

Kalmykov, A. N. (1973). (The use of heritability in selection of Trotters.) *Uchenye Zapiski Kazanskogo Gosudarstvennogo Veterinarnogo Instituta*, *121,* 23-27.

Katona, O. (1979). Genetical-statistical analysis of traits in the German Trotter. *Livestock Production Science*, *6(4)*, 407-412.

Kieffer, N. M. (1975). Heritability of racing capacity in the Thoroughbred. *Proceedings of the International Symposium on Genetics and Horse Breeding. Royal Dublin Society 17-18 Sept.* 9-18.

Klemetsdal, G., Svendsen, M. & Vangen, O. (1985). Preliminary results from genetic studies of racing performance in Norwegian Trotters. *Paper, 36th Ann. Meet. Of Eur. Assoc. for Anim. Prod.*

Langlois, B. (1975). Analyse statistique et genetique des gains des Pur Sang Anglais de trios ans dans les courses plates francaises. (Statistical and genetic analysis of the earning of 3-year-old Thoroughbred in French flat races.) *Annales de Genetique et de Selection Animale*, *7(4)*, 387-408.

Langlois, B. (1980). Heritability of racing ability in Thoroughbreds - A review. *Livestock Production Science*, *7,* 591-605.

Laughlin, H. H. (1934 a). Racing capacity in the Thoroughbred horses. 1. The measure of racing capacity. *Supplementary Publication, Carnegie Institution of Washington No. 7.*

Laughlin, H. H. (1934 b). Racing capacity in the Thoroughbred horses. 1. The measure of racing capacity. *The Scientific Monthly*, *38,* 210.

Mckee, S. L. (1995). An update on racing fatalities in the UK. *Equine Vet. Educ.*, *7,* 202-204.

Molina, A., Valera, M., Santos, D. R. & Rodero, A. (1999). Genetic parameters of morphofunctional traits in Andalusian horse. *Livestock Production Science*, *60,* 295-303.

More O'Ferrall, G. J. & Cunningham, E. P. (1974). Heritability of racing performance in Thoroughbred horses. *Livestock Production Science*, *1,* 87-97.

Mota, M. D. S. (2006). Genetic correlations between performance at different racing distances in Thoroughbreds. *Livestock Science*, *104,* 227-232.

Mota, M. D. S. & Oliveira, H. N. (2000). Precocity and performance in Brazilian Thoroughbred racehorses. *Arch. Zootec., 49,* 497-500.

Mota, M. D. S., Abrahao, A. R. & Oliveira, H. N. (2005). Genetic and environmental parameters for racing time at different distances in Brazilian Thoroughbreds. *J. Anim. Breed. Genet., 122,* 393-399.

Neisser, E. (1976). Ergebniss der prufung verschiedener Kriterien zur Bestimmung des Leistungsvermogens beim Englischen Vollblutpferd. (Evaluation of several criteria to measure performance potential in the Thoroughbred.) Leistungsprufungen von Sportpferden. 2. Internationales Wissenschaftliches Symposium (Vortage). Leipzig, am 13. und 14. Juli 1976. Leipzig, German Democratic Republic; Sektion Tierproduktion und Veterinarmedizin der Karl-Marx-Universitat. 144-154.

Neisser, E. & Schwark, H. J. (1979). Die Eignung von Rennleistungsergebnissen zur zuchterischen Bewertung von Englischen Vollbluthengsten. (Suitability of racing results for the assessment of breeding value in English Thoroughbred stallions.) *Wissenschaftliche Zeitschrift der Karl-Marx-Universitat Leipzig, Mathematisch-Naturwissenschaftliche Reihe, 28(3)*, 257-263.

Oki, H. & Sasaki, Y. (1995). *Estimation of genetic trend in racing time of Thoroughbred horses in Japan.* Equine Research Institute, Setagaya-Ku, Tokyo, *vol. 154*.

Oki, H., Sasaki, Y. & Willham, R. L. (1995). Genetic parameter estimates for racing time by restricted maximum likelihood in the Thoroughbred horse of Japan. *J. Anim. Breed. Genet., 112,* 146-150.

Oki, H., Sasaki, Y. & Willham, R. L. (1997). Estimation of genetic correlation between racing times recorded at different racing distances by restricted maximum likelihood in Thoroughbred horses. *J. Anim. Breed. Genet., 114,* 185-189.

Outram, A. K., Stear, N. A., Bendrey, R., Olsen, S., Kasparov, A., Zaibert, V., Thorpe, N. & Evershed, R. P. (2009). The Earliest Horse Harnessing and Milking. *Science, 323,* 1332-1335.

Pearson, E. S. & Hartley, H. O. (1972). *Biometrika Tables for Statisticians. Vol. 2.* Cambridge University Press, UK.

Pern, E. (1973). Prospects of improving the Thoroughbred horse. *Konevodstvo i Konnyi Sport, 5,* 15-17.

Pern, E. M. (1970). L'heritabilite de la vitesse chez le Pur Sang Anglais. *Genetika, 6,* 110-114.

Pern, E. M. (1971). Variability and heritability of speed in Thoroughbred horses. *Nauchnyi Trudy. Vsesoyuznyi Nauchno Issledovatel'skii Institut Konevodstva, 25,* 98-103.

Pirri, J. Jr. & Steele, D. G. (1952). Heritability of racing capacity in Thoroughbreds. *Journal of Animal Science, 10,* 1029.

Pirri, J. Jr. & Steele, D. G. (1952). The heritability of racing capacity. *The Blood Horse, 63,* 976.

Ricard, A. (1998). Developments in the genetic evaluation of performance traits in horses. *In 6th World Congress on Genetics Applied to Livestock Production, Armidale, Australia*, 388-395.

Robertson, W. (1971). Explanation of performance rates. *The Thoroughbred Record*. 1132-1133.

Robinson, R. A. & Gordon, B. (1988). American Association of Equine Practitioners track breakdown studies - horse results. *7th International Conference of racing Analysis and Veterinarians, Lexinton, KY, USA.* 385-394.

Rossdale, P. D., Hopes, R., Wingfield Digby, N. J. & Offord, K. (1985). Epidemiological study of wastage among racehorses 1982 and 1983. *Vet. Rec., 116,* 66-69.

Saastamoinen, M. T. (1991) Factors affecting age at onest of breaking, training, qualifying and first start Finnish Trotters. *Acta. Agric. Scand., 41,* 137-145.

Saastamoinen, M. T. & Barrey, E. (2000). Genetics of Conformation, Locomotion and Physiological Traits. In A. T. Bowling, & A. Ruvinsky (Eds.), *The Genetics of the horse* (439-472). Oxfordshire: CABI Publisher.

Saastamoinen, M. T. & Nylander, A. (1996 a). Genetic and phenotypic parameters for age at starting to race and racing performance during early career in trotters. *Livestock Production Science, 45,* 63-68.

Saastamoinen, M. T. & Nylander, A. (1996 b). Genetic and phenotypic parameters for age and speed at the beginning of racing career in trotters. *Acta. Agric. Scand. Section A. Animal Science, 46,* 39-45.

Saastamoinen, M. T. & Ojala, M. J. (1991). Influence of birth-month on age at first start and racing performance in young Trotters. *J. Agric. Sci. Finland, 41,* 437-445.

Saastamoinen, M. T., Suontama, M. & Ojala, M. (1998). Heritability of conformation traits and their relationships to racing performance in the Finnhorse Trotter. *Proceedings of the 6th World Congress on Genetic Applied to Livestock Production, Armidale, Australia,* 396-399.

Schwark, H. J. & Neisser, E. (1971). Die Zucht des Englischen Vollblutpferdes in der D. D. R. -II. Ergebnisse der Heritabilitats und Zuchtwertschatzung. *Ark. Tierzucht., 14,* 69-76.

Smith, A. M., Staniar, B. & Splan, R.K. (2006). Associations between yearling body measurements and career racing performance in Thoroughbred racehorses. *Journal of Equine Veterinary Science, 26(5),* 212-214.

Sobczynska, M. & Lukaszewicz, M. (2003). Heritability of racing merit of Arab horses. *Animal Science Papers and reports,* 21 (4), 233-239.

Sobczynska, M. & Lukaszewicz, M. (2004). Genetic parameters of racing merit of Thoroughbred horses in Poland. *J. Anim. Breed. Genet., 121,* 302-306.

Sobczynska, M. (2006). Genetic correlations between racing performance at different racing distances in Thoroughbreds and Arab horses. *Czech J. Anim. Sci., 51(12),* 523-528.

Staniar, W. B.; Kronfeld, D.S., Treiber, K.H., Splan, R.K. & Harris, P.A. (2004). Growth rate consists of baseline and systematic deviation components in Thoroughbreds. *J. Anim. Sci., 82,* 1007-1015.

Stashak, T. S. (2002). Lameness. In Adams' lameness in horses. (5th Ed.). *Lippincott Williams and Wilkins, Philadelphia, PA, USA.*

Steele, D. G. (1944). A genetic analysis of recent Thoroughbreds, Standardbreds, and American Saddle Horses. *Bulletin, Kentucky Agricultural Experiment Station, 462.*

Stone, C, P. (1935). Sex differences in the running ability of Thoroughbred horses. *Journal of Comparative Psychology, 19,* 59.

Svobodova, S., Blouin, C. & Langlois, B. (2005). Estimation of genetic parameters of Thoroughbred racing performance in the Czech Republic. *Anim. Res., 54,* 499-509.

Taveira, R. Z., Mota, M. D. S. & Oliveria, H. N. (2004). Population parameters in Brazilian Thoroughbred. *J. Anim. Breed. Genet., 121,* 384-391.

Thiruvenkadan, A.K., Kandasamy, N. & Panneerselvam S. (2009a). Inheritance of racing performance of Thoroughbred horses. *Livestock Science*, *121,* 308-326.

Thiruvenkadan, A.K., Kandasamy, N. & Panneerselvam, S. (2009b). Inheritance of racing performance of trotter horses: an overview. *Livestock Science*, *124,* 163-187.

Tolley, E. A., Notter, D. R. & Marlowe T. J. (1985). A review of the inheritance of racing performance in horses. *Animal breeding abstracts*, *53,* 163-185.

Wallin, L., Strandberg, E. & Philipsson, J. (2001). Phenotypic relationship between test results of Swedish Warmblood horses as 4-year-olds and longevity. *Livestock Production Science*, *6,* 97-105.

Watanabe, Y. (1969). Timing as a measure of selection in Thoroughbred breeding. *Japanese Journal of Zootechnical Science*, *40,* 271-276.

Watanabe, Y. (1974). Performance rates of Thoroughbreds as a criterion of racing ability. *Jap. J. Zootech. Sci.*, *45*, 408-411.

Watanabe, Y. et al. (1966). (Cited in Langlois, 1980).

Williamson, S. A. & Beilharz, R. G. (1996). Heritabilities of racing performance in Thoroughbreds: a study of Australian data. *J. Anim. Breed. Genet.*, *113,* 505-524.

Williamson, S. A. & Beilharz, R. G. (1998). The inheritance of speed, stamina and other racing performance characters in the Australian Thoroughbred. *J. Anim. Breed. Genet.*, *115*, 1-16.

Xenophon, (369 BC). On horsemanship. In the whole works of Xenophon. Translated by Ashley, Cooper, Spelman, Smith, and Fielding (1955). New York; Bangs Brother and Co. (Cited in Tolley et al., 1985).

Reviewed by Bertrand Langlois
INRA, Department of Quantitative and Applied Genetics
78352 Jouy-en-Josas Cedex, France

In: Horses: Biology, Domestication, and Human Interactions
Editor: James E. Leffhalm
ISBN: 978-1-61761-303-6

Chapter 3

SEXUAL INTERFERENCE BEHAVIOUR BY SORRAIA MARES (*EQUUS CABALLUS*)

Filipa Heitor*[*], *Sónia Emídio and Luís Vicente
Centro de Biologia Ambiental, Faculdade de Ciências da Universidade de Lisboa, Portugal

ABSTRACT

The functions of sexual interference behaviour were studied in a managed herd of Sorraia horses. We assessed the influence of age, kinship, affiliative relationships, dominance relationships, aggressiveness, oestrus state, and reproductive state on interference patterns to test five functional hypotheses: Sexual competition, mare protection, protection of a bond, dominance, and revanchism. Sexual interference behaviours consisted mainly of agonistic interactions directed by mares to the stallion, causing interruption of the sexual behaviour. On average, 29.3% of mating attempts suffered interference. Current reproductive state seemed to be the most important factor affecting rates of interference performed by mares: Barren mares interfered at higher rates than mares that foaled. Sexual interference decreased success of mating attempts, but neither decreased foaling success of receivers nor increased foaling success of interferers in the following year. The mare protection hypothesis received the strongest support from the data, as we found evidence that mares perceived mating as a threat and interfered to protect genetically related females that were involved. However, no single hypothesis tested was completely supported by the data. Future studies should test alternative hypotheses for the function of sexual interference behaviour as well as assess the effect of reproductive management practices on interference rates.

Keywords: Horse; Sexual interference; Agonistic interaction; Reproductive state

[*] Corresponding author: Edifício C2, 1749-016, Lisboa, Portugal, E-mail: heitor.filipa@gmail.com, Phone: +351 217524311

INTRODUCTION

Sexual interference behaviours occur in several taxa and consist of social interactions directed by an animal to one or both individuals of a pair involved in sexual behaviours. Interferences by adult females in polygynous fish (Lutnesky & Kosaki, 1995), salamanders (Sparreboom, Xie, & Fei, 2001), lekking birds (Owens, Burk, & Thompson, 1994; Trail, 1990), and antelopes (Bro-Jørgensen, 2002) are believed to represent a form of intrasexual competition. Interferences in sexual behaviours also occur in a variety of primate species, but their functions are less clear (Bruce & Estep, 1992; Drukker, Nieuwenhuijsen, van der Werff ten Bosch, van Hooff, & Slob, 1991; Hohmann & Fruth, 2003).

Equids interfere in affiliative and agonistic interactions, as well as sexual behaviours, and it has been suggested that interference behaviour functions to preserve or alter relationships within social groups (Schilder, 1990; Van Dierendonck et al., 2009). Sexual interferences among horses (*Equus caballus*) are performed especially by mares and have been occasionally observed by several authors (e.g., Berger, 1986; Bristol, 1982; Monard, Duncan, & Boy, 1996; Steinbjörnsson, 2002; Tyler, 1972). Some interference behaviours seemed to be aimed at protecting female associates from stallion sexual harassment (Berger, 1986) or preventing mating between daughters and close kin males (Monard et al., 1996). Sexual interferences have also been reported in Przewalski horses (Boyd, 1991; Kolter & Zimmermann, 1988; Teixeira, 2002), feral asses (Moehlman, 1998), and domestic donkeys (Henry et al., 1991). However, to our knowledge, plains zebra (*Equus quagga*) is the only equid species in which the functions of sexual interference behaviour were studied in detail (Schilder, 1990).

The aim of this study is to investigate the functions of sexual interference behaviours performed by Sorraia mares and to assess their impact on reproductive success. This study is especially relevant for management of the Sorraia horse, a Portuguese autochthonous breed which is considered to be at critical-maintained risk status by FAO (Scherf, 2000) and shows high rates of sexual interference. We tested five non-mutually exclusive functional hypotheses which are described below, together with their possible underlying proximate causes and predictions (P). Each prediction was assigned a combination of letters and numbers that represent the hypothesis and prediction number, as follows.

Sexual Competition (SC)

The social and breeding unit of feral horse populations is the band, a stable group composed of one or more stallions, several adult mares and their offspring (Kaseda, Khalil, & Ogawa, 1995; Linklater, Cameron, Minot, & Stafford, 1999; Tyler, 1972). If sexual interference is a form of intrasexual competition among mares for mating access to the stallion, interferers should attempt to stop mating or to replace the mare involved in the sexual behaviour and mate with the stallion, to increase their own reproductive success or decrease that of others. Interference behaviours should consist of agonistic interactions directed to the mare (SC-P1) or sexual behaviours directed to the stallion (SC-P2). Interferences could be performed more frequently by either of the following: Older mares, because as residual reproductive value declines with age (Cameron, Linklater, Stafford, & Minot, 2000) older mares should strive for opportunities to reproduce (SC-P3), oestrous mares (SC-P4), or barren mares (SC-P5). Interferences could be directed more frequently to mating attempts involving either of the following: Mares that are less genetically related to the interferer, according to kin selection theory (Hamilton, 1964) (SC-P6), oestrous mares (SC-P7), sexually receptive mares (SC-P8), or mares with a foal (SC-P9). The short-term consequences of interference should consist of: Interference decreases success of mating attempts (SC-P10), stallion attempts mating with the interferer immediately after interference (SC-P11). The long-term consequences could be one of the following: Higher mating frequency or success (SC-P12) and higher probability of foaling (SC-P13) of mares that perform higher rates of interference, lower mating frequency or success (SC-P14) and lower probability of foaling (SC-P15) of mares that receive higher rates of interference, mares that perform higher rates of interference have shorter time to conception (SC-P16) and their foals are born earlier (SC-P17), mares that receive higher rates of interference have longer time to conception (SC-P18) and their foals are born later (SC-P19). Mares could benefit from having their foals born earlier in the following year if older foals achieve higher rank (SC-P20), as reported by some authors (Weeks, Crowell-Davis, Caudle, & Heusner, 2000; Wells & von Goldschmidt-Rothschild, 1979).

Mare Protection (MP)

Sexual harassment by stallions can impose reproductive costs to mares (Berger, 1983; Kaseda et al., 1995; Linklater et al., 1999). Interferences could function to protect a mare from sexual behaviours, because these are perceived as threats to the mare involved. Interference behaviours should then consist of agonistic interactions directed to the stallion (MP-P1), especially when he is involved in mating attempts with unreceptive mares (MP-P2). Interferences could be directed more frequently to mating attempts involving one of the following: Mares from whom protection through interference is received more frequently (relative reciprocity *sensu* Hemelrijk, 1990a) (MP-P3), close kin (Hamilton, 1964) (MP-P4), or female associates (MP-P5).

Protection of a Bond (PB)

Bonds among horses contribute to stable group membership, which provides benefits to mares in terms of increased reproductive success (Kaseda et al., 1995; Linklater et al., 1999). Horses (Van Dierendonck et al., 2009) and zebras (Schilder, 1990) perform interference in affiliative interactions to preserve social bonds. Similarly, sexual interference could function to prevent disruption of affiliative relationships between the interferer and one or both individuals involved in sexual behaviours, because interferers perceive sexual behaviours as threats to those relationships. If this were the case, interference behaviours should consist of agonistic interactions directed to non-associates (PB-P1) or affiliative interactions directed to associates (PB-P2). In addition, interferences should be directed more frequently to mating attempts involving individuals with whom the interferer has developed stronger affiliative relationships (PB-P3).

Dominance (D)

Dominance relationships reduce aggression between group-living animals by resolving conflicts of interest over access to resources in a ritualized way (Kaufmann, 1983; Richards, 1974; Van Doorn, Hengeveld, & Weissing, 2003) and these relationships are maintained largely by submissive behaviours of subordinates (Kaufmann, 1983). However, dominance relationships within a group may not all be equally stable and clear. Interference could be a way for dominants to assert their dominance status relative to one or both individuals involved in the sexual behaviour. Interference behaviours should then consist of agonistic interactions (D-P1) and should be directed more frequently to mating attempts involving subordinates (D-P2) and individuals with whom dominance relationships are less clear (D-P3). Moreover, patterns of interference could be similar to general patterns of given agonistic interactions (D-P4).

Revanchism (R)

High-ranking mares may benefit from sexual preference by the stallion (Asa, Goldfoot, & Ginther, 1979), low aggression directed to them and their offspring (Rutberg & Greenberg, 1990; Weeks et al., 2000), and higher quality diets which allow them to produce faster growing offspring (Duncan, 1992). Therefore, mares could be expected to strive for opportunities to increase their dominance rank. Interference could be a way for subordinates to contest their dominance status relative to one or both individuals involved in the sexual behaviour. In addition, interference could be a form of retaliation for aggression received from dominants at a time when they are most vulnerable and least likely to retaliate. We make the following predictions: Interference behaviours should consist of agonistic interactions (R-P1), interferences should be directed more frequently to mating attempts involving dominants (R-P2) and individuals that are more aggressive towards the interferer (R-P3), and receivers of interference should be unlikely to respond aggressively (R-P4).

MATERIAL AND METHODS

Subjects and Study Site

The Sorraia horse is believed to be a direct descendant from the primitive South Iberian horse (Luís, 2006). The population presents high inbreeding levels and currently includes less than 200 individuals, of which less than 100 are brood mares. We collected data on one of the largest subpopulations of Sorraia horses, which is kept at Alter studfarm (39°11'N, 7°39'W), Portugal, in 2001 and from 2003 to 2005. During the study period, the herd comprised 12 different adult mares, 9 to 11 mares per year. Mares were 4 to 22 years old and had known each other since birth. Nineteen foals were born (one to six foals per year), of which five died within a few hours or days after birth. Another foal was stillborn and the mother died a few weeks after the miscarriage. Foals were weaned and removed from the group by caretakers around nine months of age. Females later returned to the group at three years old, when they were sexually mature.

Pasture breeding was implemented since 1998. Each year a stallion was released with the mares and remained in the group from around February to June for reproduction. Two stallions experienced in pasture breeding were used: One between 2001 and 2004, and another in 2005. The stallions were unacquainted with the mares when they first entered the group (2001 for the first stallion and 2005 for the second).

The herd was kept on two 5.5- and 17.2-ha enclosed pastures surrounded by 1.5-m high wire fences and walls. Horses fed mainly on grasses and supplemental feed was provided in summer. Water was available from temporary water lines, watering-ponds, and canals connected to reservoirs. Horses had access to veterinary care when necessary. A few times per year, the herd was rounded up and taken to a paddock for pregnancy detection, foal removal, branding, tail and mane shearing, and collection of blood samples for paternity analyses.

Data Collection

This study was part of a broader research project on social relationships of Sorraia horses. For this reason, the methods used for data collection suffered a few changes during the study period. Data collection methods and animal treatment comply with the policy on animal ethics. Observations were conducted on foot from a 3 - 10 m distance to the horses, between 07:30 h and 16:30 h solar time in 2001, and between 06:30 h and 18:30 h solar time from 2003 to 2005. Horses were usually sampled one at a time in random order for 25-min periods by focal and instantaneous sampling (Altmann, 1974). In 2001, however, we observed up to three focal animals simultaneously when 1-2 horses were close enough to the initially assigned focal animal. If during sampling they moved apart, the previously assigned focal animal was followed. Focal animals were all adult horses in the group during 2001 (813 h of focal sampling time over 318 h of observation) and 2003 (436.5 h of focal sampling over 610.5 h of observation). In 2005, the stallion was the only adult focal animal (42 h of focal sampling over 230.5 h of observation). *Ad libitum* data were collected in 2004 (226.5 h) and throughout the study. We analysed focal and *ad libitum* data as a whole, due to the low frequencies of sexual behaviours and interference behaviours.

During focal samples, we recorded the initiator and receiver(s), reaction of the receiver(s), and context of affiliative interactions, agonistic-related interactions and sexual behaviours. Affiliative interactions included approaches (across one body-length distance), following, mutual grooming, and friendly contacts (body contacts made with ears forward or laterally positioned). Agonistic-related interactions consisted of displacements, mild threats, bite threats, bites, kick threats, kicks, chases, attacks, and herding (see Heitor, Oom, & Vicente, 2006a, for a description). Sexual behaviours comprised sexual approach, follow, body contacts, presenting to the stallion, mounting, and copulation. Copulation was recorded when the stallion gained intromission, performed several pelvic thrusts, remained stationary on the mare for a short period of time, and finally dismounted with loss of erection (Bristol, 1982).

In instantaneous samples, the identity of the focal animal's associates (horses within one body-length) was recorded at 5-min intervals to estimate the mean time that each two mares spent as associates (i.e., association index). From 2003 to 2005, maps showing relative positions of all horses were made at 1.5-h intervals, which were used to estimate association indices. We compared the estimates of association based on instantaneous samples with those extracted from maps for the 2003 and 2005 data. As these estimates were significantly and positively correlated with each other, we considered that estimates based on instantaneous samples from the 2003 data were comparable with estimates based on maps from the 2004 and 2005 data.

Data Analysis

A case of interference was defined as one or more interactions (i.e., interference behaviours) performed by a mare towards one or both individuals involved in a mating attempt (i.e., sequence composed of one or more sexual behaviours). One mating attempt may receive several interferences from different mares or several interference behaviours from the same mare. We computed the rate of interference received as the percentage of mating attempts involving a mare that suffered interference. The rate of interference given by a mare was calculated as the percentage of mating attempts involving other mares during which she interfered.

A mare was considered to be in oestrus on a given day if at least one of the following situations was observed: (1) mare displayed sexual behaviours at least once that day; (2) copulation; (3) when receiving sexual behaviours from the stallion, mare remained calm, with raised tail and one of the following behavioural patterns: (i) straddled hindlegs, (ii) clitoral winking, and (iii) voiding small amounts of urine. If no copulation was observed and the mare reacted aggressively or moved away when sexually stimulated by the stallion, we considered that she was not in oestrus (Asa et al., 1979; Waring, 1983). When a mare's oestrus state on a given day was undetermined because no sexual behaviours were observed, the following additional criteria were taken into account: (1) intervals of one to three days between oestrous days (decided on the basis of the previous criteria) were also considered to be oestrous days, and (2) mares were not in oestrus when pregnant (Asa, Goldfoot, & Ginther, 1983). Conception date was estimated by subtracting the average gestation period of 336 days (Fraser, 1992) from foal birth date.

Sexual receptivity displayed by the mare involved in each mating attempt was assessed by her reactions: Positive (calm, raised tail, straddled hindlegs, winking, urination), negative (aggression, moving away), ambivalent (combination of positive and negative signs), and neutral (all other reactions, e.g., look at the stallion, vocalizations). A Chi-Square Test of Independence using a simulation procedure (Estabrook, Almada, Almada, & Robalo, 2002) was used to compare the reactions of mares involved in mating attempts that received interference with those that were not interfered.

Dominance was assessed by the distribution of agonistic-related interactions. Kicks and kick threats were excluded because these are defensive interactions which are not appropriate to assess dominance relationships (Van Dierendonck, De Vries, & Schilder, 1995; Wells & von Goldschmidt-Rothschild, 1979). Herding was also excluded because it was performed only by the stallion and seldom occurred. Within a dyad, the horse that initiated the majority of agonistic-related interactions that caused a submissive response was considered to be dominant. Linearity of mare dominance relationships was significant (Heitor et al., 2006a; Heitor & Vicente, 2010), as tested by means of Kendall's Coefficient of Consistence *K* (Appleby, 1983; De Vries, 1995). Therefore, mares were ordered in a linear hierarchy by the 'I and SI' method (De Vries, 1998) and were assigned an ordinal rank that increases from the bottom to the top of the hierarchy.

Associations between interference, characteristics of mating attempts and short-term consequences were tested by Chi-Square Tests of Independence. Spearman's Rank Correlation Coefficient was used to analyse the relationship between rates of interference, mare age, and relationships with the stallion.

K_r Matrix Correlation Tests (Hemelrijk, 1990a) were run between matrices of interference and test matrices to assess relative reciprocity and to test the influence of relationships among mares on patterns of interference. We converted proportions into whole numbers and constructed the following square test matrices, with mares in row and column labels:

1. Kinship: kinship coefficient (based on pedigree data up to the founder animals).
2. Dominance: dominance relationships (binary matrix: 1 – row individual dominant to column individual; 0 – the reverse) and clearness of relationships. Relationships were considered clearer in dyads with higher asymmetry of agonistic-related interactions. Asymmetry was measured as: $|f_{AB}-f_{BA}|/(f_{AB}+f_{BA})$ where f_{AB} – no. interactions directed from A to B (Van Dierendonck et al., 1995).
3. Affiliative relationships: association index.
4. Aggressiveness (i.e., frequency of agonistic interactions) given and received.

Rectangular matrices of rates of interference, frequency of mating attempts (during oestrus time), mating success (i.e., percentage mating attempts with copulation during oestrus), and foal birth order (first foal to be born receives order rank 1) were created with year in row labels and mares in column labels. These matrices had missing values because group composition changed between years and some mares were barren each year. Missing values were plugged with a constant number and dummy matrices of missing values (Hemelrijk, 1990b) were also created. Then K_r Matrix Partial Correlation Tests were run between rates of interference and the other three measures, where the dummy matrices of missing values were statistically partialled out (Hemelrijk, 1990b).

Rates of interference of mares that conceived in the first oestrus period were compared with rates of interference of mares that conceived in later oestrus periods by the Mann-Whitney *U*-test. Mann-Whitney *U*-tests were also run to compare rates of interference within a year between mares that were in different reproductive states or had different dominance status relative to the stallion. Rates of interference by the same mares in different conditions were compared by Wilcoxon Paired-Sample Tests. Mares whose foals died soon after birth were treated as barren when analysing the influence of current year's reproductive state. However, these mares were considered as having foaled when analysing reproductive state in the following year because we were interested in examining the influence of interference on probability of foaling, independent from foal survival.

Fisher Combination Test (Sokal & Rohlf, 1995) was used to combine the probabilities of statistical tests that were applied separately for each year of data.

The significance level was set at .05, but some trends were considered at an alpha level of .1. Statistical tests were one-tailed when testing significance in a particular direction and two-tailed when testing significance in both directions. In one-tailed tests, right tail probabilities, p_r, or left tail probabilities, p_l , are reported, depending on the predicted direction of significance. Matrix Correlation Tests were run on MatrixTester v2.2.3b © by Hemelrijk and probabilities were based on 2000 permutations. Standard statistical tests were performed on Statistica 5.1 © StatSoft, Inc. and StatXact 8.0 © Cytel, Inc. Chi-Square Tests of Independence using 10 000 simulations were performed on ACTUS2 © by Estabrook and Estabrook.

Results

The predictions of each hypothesis and the corresponding results are summarized in Table 1. Because some hypotheses have common predictions, the test results for each prediction will be presented according to the type of pattern of sexual interference and not according to the hypothesis where the prediction is included. Each prediction will be identified by the respective combination of letters and numbers. In the Discussion section, the results concerning the predictions of each hypothesis will be globally discussed.

Characteristics of Sexual Interference Behaviours

The mean percentage of mating attempts that received interference per year was 29.3 ± 6.4% ($M \pm SD$; range per year: 21.6 - 36.8%). Every mare performed and received interferences at least once during the study period and occasionally several mares interfered simultaneously.

Type of interference behaviour and receivers of interference (SC-P1, P2; MP-P1; PB-P1, P2; D-P1; R-P1)

Out of 222 interferences over the four years, 184 were aimed only at the stallion ($M \pm SD$: 81.2 ± 6.0% per year), 37 were directed to both individuals of the pair and one was directed to the female. The greatest percentage of interference behaviours were agonistic-related interactions ($M \pm SD$: 72.6 ± 5.4%; range per year: 68.9 – 80.6%), mainly threats. The remaining interference behaviours were affiliative interactions and only up to two interference behaviours per year consisted of sexual behaviours.

Reciprocity of interference (MP-P3)

Sexual interferences were not reciprocal: Mares did not direct interferences more frequently to mares that interfered in their sexual behaviours more often (K_r Matrix Correlation Test: 2001: $K_r = 31$, $n = 10$, $p_r = .12$; 2003: $K_r = -20$, $n = 9$, $p_r = .86$; 2004: $K_r = -5$,

$n = 11, p_r = .61$; 2005: $K_r = -16, n = 10, p_r = .85$; Fisher Combination Test: $\chi^2 = 5.86, df = 8, p > .5$).

Influence of Individual and Social Factors on Rates of Interference

Older mares perform higher rates of interference (SC-P4)

Older mares did not interfere more often than younger mares, as indicated by the non-significant correlation between median age and median interference rate throughout the study period (Spearman's Rank Correlation Coefficient: $r_s = -.38, n = 12, p = .89$).

Interference and kinship (SC-P6; MP-P4)

Sexual interferences were directed significantly more often to females that were more genetically related to the interferer (Table 2).

Interference is given more often to associates (MP-P5; PB-P3)

Rate of interference performed tended to be positively correlated with affiliative relationships among mares and with affiliative relationships between the stallion and female interferers, but not significantly so (Tables 2 and 3).

Interference and dominance relationships (D-P2; R-P2)

Rates of interference given to dominant mares were not significantly different from rates of interference given to subordinate mares (Table 2). Rates of interference performed also did not significantly differ between dominants and subordinates to the stallion (Mann-Whitney *U*-test: 2001: $U = 8, n_D = 8, n_S = 2, p = .56$; 2003: $U = 10, n_D = 7, n_S = 2, p = .25$; 2005: $U = 16, n_D = 7, n_S = 3, p = .13$; Fisher Combination Test: $\chi^2 = 8.01, df = 6, p > .1$).

Interference is given more frequently to individuals with whom dominance relationships are less clear (D-P3)

Mares interfered more frequently in mating attempts involving females with whom dominance relationships were less clear (Table 2). However, rates of interference performed were not significantly correlated with clearness of dominance relationships between interferers and the stallion (Table 3).

Patterns of interference could be similar to general patterns of given agonistic interactions (D-P4)

Rate of interference given was positively and significantly correlated with aggressiveness given to mares (Table 2). Rate of interference was also positively correlated with aggressiveness given by mares to the stallion, but not significantly so (Table 3).

Interference is given more frequently to individuals that are more aggressive towards the interferer (R-P3)

Rate of interference given to mares was not significantly correlated with aggressiveness received from those mares (Table 2). Rate of interference tended to be positively correlated with aggressiveness received by mares from the stallion, but not significantly so (Table 3).

Table 1. Summary of the predictions of the five functional hypotheses tested in this study and the corresponding results

Patterns of Sexual Interference	Sexual Competition	Mare Protection	Protection of a Bond	Dominance	Revanchism
Type of Interference Behaviour	Sexual –; Agonistic*	Agonistic*	Affiliative –; Agonistic*	Agonistic*	Agonistic*
Receiver of Interference Behaviour	Stallion*; Mare –	Stallion*			
Reciprocity of Interference		Reciprocal –			
Interference Performed by Mares that are:					
Age	Older –				
Oestrus State	Oestrous –				
Reproductive State	Barren*				
Aggressiveness				More Aggressive*	
Interference Given to Mares that are:					
Kinship	Less Related –	More Related*			
Oestrus State	Oestrous*				
Sexual Receptivity	Positive –	Negative –			
Current Year's Reproductive State	Foaled +				
Dominance				Subordinate –	Dominant –
Clearness of Dominance Relationships				Less Clear*	
Affiliative Relationships		Associates +	Associates +		
Aggressiveness					More Aggressive –
Short-Term Consequences:					
Success of mating attempts	Lower*				
Reaction of receivers of interference	Stallion: Sexual –				Non-Agonistic*
Long-Term Consequences (Initiators; Receivers):					
Frequency of Mating Attempts	Higher*; Lower –				
Mating Success	Higher –; Lower –				
Following Year's Reproductive State	Foaled –; Barren –				
Time to Conception	Shorter –; Longer –				
Foal Birth Order in the Following Year	Earlier –; Later –				
Correlation Foal Birth Order and Rank	Negative –				

* prediction supported ; – prediction not supported; + tendencies in the predicted direction or prediction partially supported

Table 2. Influence of Relationships Among Mares on Rates of Interference Given, as Assessed by the K_r Matrix Correlation Test (2001, 2005: *n* = 10; 2003: *n* = 9; 2004: *n* = 11) and the Fisher Combination Test (*df* = 8 when Combining Four Tests, and *df* = 4 when Combining Two Tests)

	K_r Tests				Fisher Test	
Measure	2001	2003	2004	2005	χ^2	p [a]
Kinship	71*	45*	21	-17	19.90	$p < .025$
Affiliative Relationships	59*	29	-7	-34*	14.28	$p_r < .1$
Dominance Relationships	16	9	31*	27	14.01	$p < .1$
Clearness of Dominance	-55**	3	-	-	10.02	$p_l < .05$
Aggressiveness Given	33	26	-	-	9.98	$p_r < .05$
Aggressiveness Received	5	-22	-	-	1.97	$p_r > .5$

* $p < .05$.
** $p < .01$.
[a] *p*-values of Fisher Combination Test: p – two-tailed probability, p_r - one-tailed probability in the right tail of the permutation distribution, p_l - one-tailed probability in the left tail of the permutation distribution.

Table 3. Influence of Relationships Between the Stallion and the Mares on Rates of Interference Given, as Measured by the Spearman's Rank Correlation Coefficient (2001, 2005: *n* = 10; 2003: *n* = 9; 2004: *n* = 11) and the Fisher Combination Test (*df* = 8 when Combining Four Tests, and *df* = 6 when Combining Three Tests)

	r_s Tests				Fisher Test	
Measure	2001	2003	2004	2005	χ^2	p [a]
Affiliative Relationships	- .07	.01	.59*	.42	13.61	$p_r < .1$
Clearness of Dominance	- .07	.42	-	- .30	5.17	$p_l > .5$
Aggressiveness Given by Mares	.52	.12	-	.42	11.66	$p_r < .1$
Aggressiveness Received by Mares	.45	- .10	-	.59[a]	12.15	$p_r < .1$

* $p < .05$.
[a] *p*-values of Fisher Combination Test: p_r - one-tailed probability in the right tail of the permutation distribution, p_l - one-tailed probability in the left tail of the permutation distribution.

Interference and sexual receptivity (SC-P8; MP-P2)

Mating attempts suffered interference depending on the receptivity of the mare involved (ACTUS: $\chi^2 = 8.4$, $p = .03$). However, neither mating attempts with positive reactions nor mating attempts with negative reactions were interfered significantly more often than expected.

Oestrous mares perform higher rates of interference (SC-P4)

Mean rate of interference given by mares when they were in oestrus was not higher than when they were not in oestrus (Wilcoxon Paired-Sample Test: $T = 26$, $n = 12$, $p = .17$).

Table 4. Difference between rates of interference (M ± SD) given and received by Barren Mares (B) and Mares that had a foal (F) in the current and following years, as assessed by the Mann-Whitney U-test and the Fisher combination test (df = 8 when combining four tests, and df = 6 when combining three tests)

Rate (%)	Current Year					Following Year				
	Barren	Foal	n_B	n_F	Test	Barren	Foal	n_B	n_F	Test
Given										
2001	2.5 ± 2.0	2.8	9	1	NP [a]	1.3 ± 1.2	3.4 ± 1.9	4	6	$U = 20, p = .06$
2003	5.3 ± 4.9	1.0 ± 0.5	6	3	$U = 17, p = .02$	7.4 ± 6.7	2.1 ± 1.2	3	6	$U = 5, p = .81$
2004	7.8 ± 7.6	1.1 ± 0.7	5	6	$U = 25, p = .04$	5.2 ± 7.5	3.6 ± 5.5	4	7	$U = 20, p = .66$
2005	5.2 ± 4.0	0.4 ± 0.4	6	4	$U = 21, p = .03$	4.0 ± 4.0	0.4 ± 0.5	8	2	$U = 3, p = .87$
Total					$\chi^2 = 20.90, p < .005$					$\chi^2 = 7.28, p > .50$
Received										
2001	22.7 ± 14.5	18.2	9	1	NP [a]	16.0 ± 3.7	26.5 ± 16.7	4	6	$U = 4.5, p = .94$
2003	29.3 ± 12.0	36.6 ± 12.4	6	3	$U = 14, p = .13$	19.5 ± 5.5	37.9 ± 8.8	3	6	$U = 2, p = .95$
2004	34.2 ± 10.0	42.4 ± 21.6	5	6	$U = 20, p = .21$	35.5 ± 13.1	40.5 ± 19.8	4	7	$U = 4.5, p = .62$
2005	24.5 ± 13.6	39.6 ± 11.0	6	4	$U = 21, p = .03$	27.0 ± 12.7	44.9 ± 13.6	8	2	$U = 2, p = .91$
Total					$\chi^2 = 14.03, p < .05$					$\chi^2 = 1.37, p > .99$

[a] Test not performed due to small sample size of mares that had foals.

Barren mares perform higher rates of interference (SC-P5)

Within each year, barren mares interfered significantly more often than mares with foals, as predicted (Table 4). The six mares that experienced different reproductive states interfered more frequently when they were barren than when they had a foal (Wilcoxon Paired-Sample Test: $T = 0$, $n = 6$, $p = .02$).

Interference is given more frequently to oestrous mares (SC-P7)

Mating attempts involving oestrous mares were interfered significantly more often than mating attempts involving non-oestrous mares (33.3% of 386 oestrous vs. 23.8% of 189 non-oestrous, $\chi^2 = 5.27$, $df = 1$, $p = .02$).

Interference is given more frequently to mares with foals (SC-P9)

Within each year, mares with foals received significantly higher rates of interference than barren mares, as predicted (Table 4). However, rates of interference received were not higher when mares had foals than when the same mares were barren (Wilcoxon Paired-Sample Test: $T = 7$, $n = 6$, $p = .28$).

Consequences of Sexual Interference

Short-term consequences

Interference decreases success of mating attempts (SC-P10). Interferences usually caused the stallion and the mare involved in the sexual behaviour to move away from each other and from the interferer ($M \pm SD$: 72.5 ± 6.5% per year). After receiving interference, 85.2 ± 6.3% ($M \pm SD$ per year, range = 80 - 94.1%) of mating attempts were interrupted for at least 30 s. Mating attempts that received interference were significantly less successful than those that were not interfered (2.9% success of 208 interfered vs. 17.6% success of 506 not interfered, $\chi^2 = 27.63$, $df = 1$, $p < .001$).

Stallion attempts mating with the interferer immediately after interference (SC-P11). The stallion seldom directed sexual behaviours to the interferer mare within the 30 s immediately after her interference (range = 0-4 times per year). Moreover, the stallion copulated with the interferer within 30 s after interference only once in the whole study.

Receivers of interference should be unlikely to respond aggressively (R-P4). Aggressive reactions by the receivers of interference towards the interferer were observed only once in two of the four study years.

Long-term consequences

Higher mating frequency or success of mares that perform higher rates of interference (SC-P12). Mares that performed higher rates of interference were involved significantly more often in mating attempts, but their mating success was not higher (Table 5).

Higher probability of foaling of mares that perform higher rates of interference (SC-P13). Mares that performed higher rates of interference were not more likely to foal in the following year (Table 4). In addition, mares had not performed higher rates of interference when they foaled in the following year than when they did not foal (Wilcoxon Paired-Sample Test: $T = 13$, $n = 7$, $p = .47$).

Lower mating frequency or success of mares that receive higher rates of interference (SC-P14). Rates of interference received were not significantly correlated either with frequencies of mating attempts or with mating success (Table 5).

Lower probability of foaling of mares that receive higher rates of interference (SC-P15). Rates of interference received by mares that foaled were not lower than rates of interference received by mares that did not foal in the following year. Contrary to our predictions, mares that received higher rates of interference were significantly more likely to foal (Table 4). Moreover, mares had not received lower rates of interference when they foaled in the following year than when they did not foal and, contrary to the predictions, the opposite trend was found (Wilcoxon Paired-Sample Test: $T = 5$, $n = 7$, $p = .92$).

Mares that perform higher rates of interference have shorter time to conception and their foals are born earlier (SC-P16, P17). Rate of interference given did not affect time to conception (Mann-Whitney: U-test: $U = 18$, $n_1 = 6$, $n_2 = 6$, $p = .53$) and foal birth order in the following year (Table 5).

Mares that receive higher rates of interference have longer time to conception and their foals are born later (SC-P18, P19). Mares that received higher rates of interference did not conceive later (Mann-Whitney: U-test: $U = 14$, $n_1 = 6$, $n_2 = 6$, $p = .71$) and, contrary to the prediction, their foals were born earlier (Table 5).

Foals that are born earlier achieve higher rank (SC-P20). We did not perform statistical tests to analyse the linearity of foal dominance hierarchies and the correlation between foal birth order and rank because the number of foals per year was small (3 to 6 foals between 2003 and 2005). However, foals that were born earlier were not more likely to be dominant: the oldest foal was dominant in 11 out of 24 dyads (Binomial Test: $p = .58$).

Table 5. Relationship between rates of interference and frequency of mating attempts, mating success and foal birth order in the following year, as assessed by the K_r matrix partial correlation test

	Interference Given			Interference Received		
Measure	$\tau_{Krxy.z}$	n	p	$\tau_{Krxy.z}$	n	p
Frequency of Mating Attempts	- .45	12	.0005	.04	12	.38
Mating Success	.05	11	.38	- .10	11	.31
Birth Order	- .05	10	.39	- .49	10	.0005

Discussion

None of the five functional hypotheses tested in this study was completely supported by the data, but some hypotheses were partially supported. Results regarding the predictions of each hypothesis are discussed in detail below.

Sexual Competition Hypothesis

There was no evidence that interferers were trying to replace the mare and mate with the stallion, because they rarely directed sexual behaviours to the stallion or mated with him after interference. Interferences were mostly agonistic interactions directed to the male, not to the female, a finding which is inconsistent with this hypothesis. Interferences were also directed predominantly to the male in some studies (Bristol, 1982; Berger, 1986; Drukker et al., 1991; Schilder, 1990), but were aimed at the female in others (Hohmann & Fruth, 2003; Lutnesky & Kosaki, 1995; Owens et al., 1994; Trail, 1990; Tyler, 1972).

Mares did not interfere more frequently during oestrus than during non-oestrus periods. However, consistent with the predictions, barren mares performed higher rates of interference than mares with foals. Moreover, mating attempts involving oestrous mares and mares with foals were preferentially interfered. Female stumptail macaques also interfere more often during normal ovarian cycles than when they are pregnant or lactating (Drukker et al., 1991). Similar to these findings, the frequency of interference by female angelfish, *Centropyge potteri*, is highest before spawning and almost negligible thereafter (Lutnesky & Kosaki, 1995).

The percentage of mating attempts that received interference (29.3%) was lower among Sorraia horses than among plains zebras (51%; Schilder, 1990). However, Sorraia mares were more successful at interrupting mating attempts in which they interfered and preventing mating than zebras (85% interruption in Sorraia mares vs. 44% in zebras). Although interferences were effective at preventing mating, at least for the short period of time considered (30 s), there was no evidence to support the predictions that interference increases the interferer's reproductive success or decreases that of others. Mares that performed higher rates of interference had higher frequency of mating attempts, as predicted. However, we found no evidence of higher mating success, increased probability of foaling, shorter time to conception, or foals being born earlier among mares that interfered more frequently. Receiving interferences decreased success of mating attempts, as predicted, but frequency of mating attempts and mating success were not lower in mares that received higher rates of interference. Contrary to the predictions of sexual competition, mares that received higher rates of interference were more likely to foal in the following year and their foals were born earlier. Finally, a correlation between birth order and foal rank was not found.

Altogether, the findings provide only partial support for the sexual competition hypothesis.

Mare Protection Hypothesis

This hypothesis was partially supported by the data. Interference behaviours consisted of aggression towards the stallion, and interferences were given more frequently to mares that were more genetically related to the interferer. This finding suggests that some form of kin discrimination may be present in horses (Heitor, Oom, & Vicente, 2006b). Nevertheless, mares did not interfere more frequently when the female was showing a negative reaction to the stallion's sexual behaviours. Furthermore, mares did not prefer to interfere in sexual behaviours involving females that interfered with their sexual behaviours more often. In summary, there is some evidence that mares could perceive mating as a threat and tried to protect genetically related females that were involved (Berger, 1986; Schilder, 1990). However, the perception of mating as a threat does not seem to depend on the reactions of the females involved.

Protection of a Bond Hypothesis

Mares that formed stronger affiliative relationships with the stallion tended to interfere more often. However, the finding that interferences consisted mainly of agonistic interactions towards the stallion does not support the hypothesis of protection of a bond with the stallion. Interferences tended to be more frequent among mares with stronger affiliative relationships, as predicted by protection of a bond with the mare, but not significantly so. These results contrast with those of Schilder (1990) and Steinbjörnsson (2002), who reported that interferers tended to be associates of the female involved in the sexual behaviour. Hence, protection of a bond does not provide a satisfactory explanation for sexual interference.

Dominance Hypothesis

Consistent with this hypothesis, interference behaviours consisted mainly of agonistic interactions. In addition, interferences were aimed preferentially at mating attempts involving mares to whom interferers directed agonistic interactions more frequently and with whom dominance relationships were less clear. Nevertheless, interference behaviours were mainly aimed at the stallion and rate of interference was not higher towards subordinates, which is inconsistent with assertion of dominance status with female subordinates. Boyd (1991), Bristol (1982), Kolter and Zimmermann (1988), and Tyler (1972) reported that interferences were performed by dominant mares towards subordinates. However, like in Sorraia horses, dominance relationships among plains zebras did not affect rates of interference performed (Schilder, 1990).

Female dominants to the stallion did not interfere more frequently than subordinates. In addition, no correlation was found between rate of interference and frequency of agonistic interactions given to the stallion or clearness of dominance relationships. Therefore, evidence that interferences were aimed at asserting dominance status with other mares or with the stallion is scant.

Revanchism Hypothesis

As predicted by this hypothesis, receivers of interference almost never reacted aggressively towards the interferer and interference behaviours consisted mainly of agonistic interactions. However, interference behaviours were mostly directed to the stallion and interferences were not directed from subordinates to dominants more frequently than the reverse. Moreover, interferences were not directed to females from whom the interferer received higher frequency of agonistic interactions. These findings are inconsistent with a possible role of interference in contests over dominance status among females.

Females that received higher frequency of agonistic interactions from the male did not perform significantly higher rates of interference, and subordinates to the stallion did not perform higher rates of interference than dominants. Therefore, sexual interference does not seem to function to contest dominance relationships neither with mares nor with the stallion.

The fact that the hypotheses were only partially supported by the data suggests that sexual interference behaviour may serve a different function not tested in this study. Nevertheless, one should be careful when interpreting these results because the small study group size may not have allowed sufficient statistical power to detect some relationships between patterns of interference and other variables. Therefore, more data needs to be collected in this herd and in other Sorraia horse groups before conclusions regarding the function of sexual interference behaviour can be drawn.

Conclusion

Some patterns were found to characterize sexual interference behaviours among Sorraia horses: Sexual interference behaviours consisted mainly of agonistic interactions directed by mares to the stallion, interference decreased success of mating attempts and barren mares interfered most frequently. Mares with foals invest large amounts of time and energy in maternal care (Berger, 1986; Cameron et al., 2000). Leaving a foal to perform sexual interferences is a risk to the foal's safety and probably imposes additional costs to mothers in terms of time and energy which could decrease their ability to care for current offspring. Therefore, barren mares probably incur lower costs from performing interference than mares with foals.

Mare protection was the hypothesis that received the strongest support from the data. Evidence was found that mares may perceive mating as a threat and interfere to protect genetically related females. Sexual behaviours could be perceived as a conflict between the stallion and a mare because the stallion was a recently introduced member in the group. Consistent with this view, sexual interference behaviours were only occasionally observed in feral horse groups, where long term stable relationships are formed between the harem stallion and mares. Therefore, the common reproductive management practice of keeping the stallion with the mares only during the breeding season may be a proximate cause of the high sexual interference rates among Sorraia horses and deserves further research.

Studies that combine behavioural data with experimental manipulation of the social environment, hormonal analyses and assessment of long term effects of interference on reproductive success and welfare could provide a greater understanding of the benefits and costs of sexual interference behaviour. Such knowledge would allow the development of management practices that minimize the frequency of sexual interference behaviours and enhance mating success. This is especially relevant for the conservation of an endangered breed such as the Sorraia horse.

ACKNOWLEDGMENTS

We thank Dr. João Costa Ferreira for permission to conduct this study at Alter studfarm and Dr. Maria do Mar Oom for providing the kinship data. Dr. Patrick Duncan provided insightful comments on an earlier version of this paper. We are also indebted to Dr. Charlotte Hemelrijk and Dr. George Estabrook, who kindly provided their computer programs and user instructions. This study was supported by the Fundação para a Ciência e a Tecnologia (FCT) through a Doctoral Fellowship (SFRH/BD/8792/2002) granted to Filipa Heitor.

REFERENCES

Altmann, J. (1974). Observational study of behavior: Sampling methods. *Behaviour, 49,* 227-265.

Appleby, M. C. (1983). The probability of linearity in hierarchies. *Animal Behaviour, 31,* 600-608.

Asa, C. S., Goldfoot, D. A. & Ginther, O. J. (1979). Socio-sexual behavior and ovulatory cycles of ponies (*Equus caballus*) observed in harem groups. *Hormones and Behavior, 13,* 49-65.

Asa, C. S., Goldfoot, D. A. & Ginther, O. J. (1983). Assessment of the sexual behaviour of pregnant mares. *Hormones and Behavior, 17,* 405-413.

Berger, J. (1983). Induced abortion and social factors in wild horses. *Nature, 303,* 59-61.

Berger, J. (1986). *Wild horses of the Great Basin.* Chicago: University of Chicago Press.

Boyd, L. E. (1991). The behavior of Przewalski's horses and its importance to their management. *Applied Animal Behaviour Science, 29,* 301-318.

Bristol, F. (1982). Breeding behaviour of a stallion at pasture with 20 mares in synchronized oestrus. *Journal of Reproduction and Fertility, 32 (Suppl.),* 71-77.

Bro-Jørgensen, J. (2002). Overt female mate competition and preference for central males in a lekking antelope [Electronic version]. *PNAS, 99,* 9290-9293.

Bruce, K. E. & Estep, D. Q. (1992). Interruption of and harassment during copulation by stumptail macaques, *Macaca arctoides*. *Animal Behaviour, 44,* 1029-1044.

Cameron, E. Z., Linklater, W. L., Stafford, K. J. & Minot, E. O. (2000). Aging and improving reproductive success in horses: Declining residual reproductive value or just older and wiser? *Behavioral Ecology and Sociobiology, 47,* 243-249.

De Vries, H. (1995). An improved test of linearity in dominance hierarchies containing unknown or tied relationships. *Animal Behaviour, 50,* 1375-1389.

De Vries, H. (1998). Finding a dominance order most consistent with a linear hierarchy: a new procedure and review. *Animal Behaviour, 55,* 827-843.

Drukker, B., Nieuwenhuijsen, K., van der Werff ten Bosch, J. J., van Hooff, J. A. R. A. M. & Slob, A. K. (1991). Harassment of sexual interactions among stumptail macaques, *Macaca arctoides*. *Animal Behaviour, 42,* 171-182.

Duncan, P. (1992). *Horses and grasses: The nutritional ecology of equids and their impact on the Camargue*. New York: Springer-Verlag.

Estabrook, G. F., Almada, V. C., Almada, F. J. & Robalo, J. I. (2002). Analysis of conditional contingency using ACTUS2 with examples from studies of animal behavior. *Acta Ethologica, 4,* 73-80.

Fraser, A. F. (1992). *The behaviour of the horse.* Wallingford: CAB International.

Ginther, O. J., Scraba, S. T. & Nuti, L. C. (1983). Pregnancy rates and sexual behaviour under pasture breeding conditions in mares. *Theriogenology, 20,* 333-345.

Hamilton, W. D. (1964). The genetical evolution of social behaviour. I. *Journal of Theoretical Biology, 7,* 1-16.

Heitor, F. & Vicente, L. (2010). Affiliative relationships among Sorraia mares: Influence of age, dominance, kinship and reproductive state. *Journal of Ethology, 28,* 133-140.

Heitor, F., Oom, M. M. & Vicente, L. (2006a). Social relationships in a herd of Sorraia horses: Part I. Correlates of social dominance and contexts of aggression. *Behavioural Processes, 73,* 170-177.

Heitor, F., Oom, M. M. & Vicente, L. (2006b). Social relationships in a herd of Sorraia horses: Part II. Factors affecting affiliative relationships and sexual behaviours. *Behavioural Processes, 73,* 231-239.

Hemelrijk, C. K. (1990a). Models of, and tests for, reciprocity, unidirectionality and other social interaction patterns at a group level. *Animal Behaviour, 39,* 1013-1029.

Hemelrijk, C. K. (1990b). A matrix partial correlation test used in investigations of reciprocity and other social interaction patterns at group level. *Journal of Theoretical Biology, 143,* 405-420.

Henry, M., McDonnell, S. M., Lodi, L. P. & Gastal, E. L. (1991). Pasture mating behaviour of donkeys (*Equus asinus*) at natural and induced oestrus. *Journal of Reproduction and Fertility, 44 (Suppl.),* 77-86.

Hohmann, G. & Fruth, B. (2003). Intra- and inter-sexual aggression by bonobos in the context of mating [Electronic version]. *Behaviour, 140,* 1389–1413.

Kaseda, Y., Khalil, A. M. & Ogawa, H. (1995). Harem stability and reproductive success of Misaki feral mares. *Equine Veterinary Journal, 27,* 368-372.

Kaufmann, J. H. (1983). On the definitions and functions of dominance and territoriality. *Biological Reviews, 58,* 1-20.

Kolter, L. & Zimmermann, W. (1988). Social behaviour of Przewalski horses (*Equus p. przewalskii*) in the Cologne Zoo and its consequences for management and housing. *Applied Animal Behaviour Science, 21,* 117-145.

Linklater, W. L., Cameron, E. Z., Minot, E. O. & Stafford, K. J. (1999). Stallion harassment and the mating system of horses. *Animal Behaviour, 58,* 295-306.

Luís, C. (2006). *The Sorraia horse, a case study of a small and closed population. Contribution of molecular markers to the genetic conservation and management of an endangered breed.* PhD Thesis, University of Lisbon, Portugal.

Lutnesky, M. M. F. & Kosaki, R. K. (1995). Female-female competition in a coral reef fish and a test of the temporal threshold model of polygynous mating. *American Naturalist, 146,* 832-847.

Moehlman, P. D. (1998). Behavioral patterns and communication in feral asses (*Equus africanus*). *Applied Animal Behaviour Science, 60,* 125-169.

Monard, A. M., Duncan, P. & Boy, V. (1996). The proximate mechanisms of natal dispersal in female horses. *Behaviour, 133,* 1095-1124.

Owens, I. P. E., Burk, T. & Thompson, D. B. A. (1994). Extraordinary sex ratio in the Eurasian dotterel: Female mating arenas, female-female competition and female mate choice. *American Naturalist, 144,* 76-100.

Richards, S. M. (1974). The concept of dominance and methods of assessment. *Animal Behaviour, 22,* 914-930.

Rutberg, A. T. & Greenberg, S. A. (1990). Dominance, aggression frequencies and modes of aggressive competition in feral pony mares. *Animal Behaviour, 40,* 322-331.

Scherf, B. D. (2000). *World watch list for domestic animal diversity* (3rd ed.). Rome: FAO.

Schilder, M. B. H. (1990). Interventions in a herd of semi-captive plains zebras. *Behaviour, 112,* 53-83.

Sokal, R. R. & Rohlf, F. J. (1995). *Biometry: The principles and practice of statistics in biological research.* New York: W.H. Freeman.

Sparreboom, M., Xie, F. & Fei, L. (2001). Reproductive behaviour of the Chinhai Salamander (*Echinotriton chinhaiensis*) (Caudata: Salamandridae) [Electronic version]. *Amphibia-Reptilia, 22,* 309-320.

Steinbjörnsson, B. (2002). Sexual behavior in herds of Icelandic horses. In S. McDonnell & D. Mills (Eds.), *Horse behavior and welfare.* A Dorothy Russell Havemeyer Foundation Workshop, June 13-16, Hólar, Iceland. Retrieved 20 September, 2004, from http://www2.vet.upenn.edu/labs/equinebehavior/hvnwkshp/ hv02/steinbj.htm.

Teixeira, C. (2002). *Time budgets e interacções sociais na adaptação de um harém de cavalos selvagens da Mongólia* (*Equus ferus przewalskii* Poliakov, 1881) *na Coudelaria de Alter.* Graduation Report, Faculty of Sciences of the University of Lisbon, Portugal.

Trail, P. W. (1990). Why should lek-breeders be monomorphic? *Evolution, 44,* 1837-1852.

Tyler, S. (1972). The behaviour and social organization of the New Forest ponies. *Animal Behaviour Monographies, 5,* 85-196.

Van Dierendonck, M. C., De Vries, H. & Schilder, M. B. H. (1995). An analysis of dominance, its behavioural parameters and possible determinants in a herd of Icelandic horses in captivity. *Netherlands Journal of Zoololgy, 45,* 362-385.

Van Dierendonck, M. C., De Vries, H., Schilder, M. B. H., Colenbrander, B., Thórhallsdóttir, A. G. & Sigurjónsdóttir, H. (2009). Interventions in social behaviour in a herd of mares and geldings. *Applied Animal Behaviour Science, 116,* 67-73.

Van Doorn, G. S., Hengeveld, G. M. & Weissing, F. J. (2003). The evolution of social dominance I: Two-player models. *Behaviour, 140,* 1305-1332.

Waring, G. H. (1983). *Horse behavior: The behavioral traits and adaptations of domestic and wild horses, including ponies.* Park Ridge, NJ: Noyes Publications.

Weeks, J. W., Crowell-Davis, S. L., Caudle, A. B. & Heusner, G. L. (2000). Aggression and social spacing in light horse (*Equus caballus*) mares and foals. *Applied Animal Behaviour Science, 68,* 319-337.

Wells, S. M. & von Goldschmidt-Rothschild, B. (1979). Social behaviour and relationships in a herd of Camargue horses. *Zeitschrift für Tierpsychologie, 49,* 363-380.

In: Horses: Biology, Domestication, and Human Interactions ISBN: 978-1-61761-303-6
Editor: James E. Leffhalm

Chapter 4

MANAGEMENT OF WILD HORSES WITH PORCINE ZONA PELLUCIDA: HISTORY, CONSEQUENCES, AND FUTURE STRATEGIES

Cassandra M.V. Nuñez
Department of Ecology and Evolutionary Biology,
Princeton University, Princeton, New Jersey

ABSTRACT

The advent of immunocontraception with porcine zona pellucida (PZP) has all but revolutionized wild horse management, providing a more humane method of population control than earlier strategies. Early studies on Assateague Island National Seashore have described it as an ideal form of fertility control in that it reduces the chance of conception to below 10%, can be delivered remotely, is reversible (after short-term use), lacks debilitating physiological side effects, cannot pass through the food chain, and shows minimal effects on social behaviors. However, recent research in other populations has revealed behavioral and physiological side effects of long-term PZP use. These results indicate that studies from one population may not necessarily be applicable to another, regardless of similarities in habitat and population structure. Careful study of the animals' demography, physiology, and behavior is necessary prior to and during treatment to ensure that a) the potential effects of PZP can be assessed accurately, and b) within managerial constraints, PZP effects are ameliorated as much as is possible. Here I explain the history of PZP use in wild horses, its side effects on the horses of Shackleford Banks, North Carolina in comparison to other populations, and offer management suggestions based upon wild horse biology and behavioral ecology, which may minimize or prevent these side effects in other populations.

INTRODUCTION

The modern horse (*Equus caballus*) thrived in North America for nearly five million years before its local extinction approximately 11,000 years ago [Weinstock et al. 2005]. In the late 15th century, horses were brought back to our continent by the Spanish and have since roamed free, forming feral herds in the United States and Canada [Weinstock et al. 2005]. Some argue that preservation of these animals is questionable given their recent history of reintroduction. Others maintain that they are an integral part of the United States' heritage and the natural history of our continent. Currently, the latter viewpoint has prevailed and wild horse preservation is assured by various acts in different parts of the country such as the Wild Free-Roaming Horse and Burrow Act of 1971 and the Shackleford Banks Wild Horses Protection Act of 1997. The extirpation of their natural predators and modern pressures such as ranching and farming all but necessitate some form of population management throughout the country, however the "patchwork quilt" of policies regarding wild horse preservation means that populations under separate jurisdictions are treated differently.

Managers have few options to control population numbers. Culling is not readily accepted by the public, and "gathers," in which animals are rounded up by helicopter and removed, divide family groups, significantly decreasing herd stability [Kirkpatrick et al. 1997]. In addition, the use of hormonal contraception (in both male and female animals) has proven problematic due to the dosage needed to achieve infertility, the subsequent changes to behavior, and the issue of possible consumption by non-target species [Turner and Kirkpatrick 1991]. A more recent alternative, immunocontraception, induces the production of antibodies against structural or functional molecules important to reproduction. These antibodies impede the normal cell biology of reproduction, reducing the likelihood of conception [Turner and Kirkpatrick 1991]. In the case of porcine zona pellucida (PZP), antibodies bind sperm receptors on the egg's surface, thereby preventing sperm attachment and fertilization [Sacco 1977]. Today, PZP is used to control reproduction in several species of free-ranging ungulates including white-tailed deer (*Odocoileus virginianus*) [McShea et al. 1997], elk (*Cervus elaphus*) [Heilmann et al. 1998] and the wild horse [Kirkpatrick 1990, Turner et al. 2002].

The majority of studies examining the effects of PZP application to wild horses stem from a long-term study on Assateague Island National Seashore, Maryland. Most of these studies have focused on the physiological effects of the vaccine [Kirkpatrick et al. 1992, Kirkpatrick et al. 1997, Turner et al. 2002]. Researchers have reported no debilitating side effects to PZP recipients and only minor ovulation failure and depressed urinary oestrogen concentrations with repeated applications [Kirkpatrick et al. 1996]. In addition, the contraceptive effects of PZP have been shown to be reversible, safe for pregnant females, and do not adversely affect the survivorship or subsequent fertility of offspring born to treated individuals [Kirkpatrick and Turner 2002].

However, recent studies on the horses living on Shackleford Banks (a barrier island off the North Carolina coast) and in the western United States, demonstrate differences in recipient behavior and physiology [Nunez et al. 2009, Nunez et al. in review, Ransom et al. 2010]. Shackleford mares receiving PZP change harem groups more often and treated horses in both areas both initiate and receive more reproductive behaviors than do untreated mares. For a gregarious species like the horse, such changes may have serious social and demographic consequences. Here I give an brief overview of wild horse biology, discuss the history of PZP use in wild horses, compare its effects in different populations, and offer management suggestions based upon the species' behavioral ecology that may prevent or ameliorate these effects.

WILD HORSE BIOLOGY AND BEHAVIORAL ECOLOGY

In wild horse societies, the harem is the core social group, consisting of usually one, but sometimes two or three harem male(s), one to several female(s), and their offspring [Feist and McCullough 1976, Rubenstein 1981; 1986, Linklater et al. 2000]. Harem groups are typically stable units, showing very few changes in composition over months or years [Klingel 1975], although this can vary among populations [Tyler 1972, Rutberg 1990]. Female loyalty to the harem male and the males' ability to retain females is paramount to maintaining this stability [Feist and McCullough 1976, Rubenstein 1981, Goodloe et al. 2000]. Decreases in harem stability have been shown to affect several aspects of mare well-being, resulting in lower overall reproductive success [Kaseda et al. 1995], less time for preferred behaviors, decreased body condition and fecundity, elevated parasite levels, and increased offspring mortality [Linklater et al. 1999].

Breeding normally occurs from March through August, with most births occurring in April and May [Crowell-Davis 2007]. Gestation lasts approximately 11-12 months [Asa 2002]. Temperate-zone equids are more strictly seasonal than are tropical-zone species. However, temperate individuals can show significant variability in reproductive timing and estrous behaviors during the fall and winter months which constitute the non-breeding season [Asa 1979, Asa et al. 1980]. Photoperiod is probably the most important cue for reproduction in temperate species [Ginther 1979; 1992], although birth peaks in the tropics suggest that nutritional factors are also important [Grubb 1981, Penzhorn 1988, Churcher 1993]. In all equids, estrus recurs until pregnancy or until the end of the breeding season [Asa 2002].

For the most part, wild horses are non-territorial, with several harems sharing both feeding and water resources [Feist and McCullough 1976, Rubenstein 1981, Cameron et al. 2003]. Given this ecology, changes in the behavior and/or physiology of individuals have the potential to affect interactions with neighboring harems and may, thereby, affect significant change at the population level. As such, understanding the potential effects of PZP on individual behavior and physiology is of broad importance to maintaining a functional population of feral horses.

PZP USE IN WILD HORSES

The management of wild horses across the U.S. is conducted for a variety of reasons. In the west, the Bureau of Land Management manages several populations in order to "establish appropriate management levels [based on] the ecological balance among wild horses and burros and the BLM's multiple use mission" [Bureau of Land Management 2010]. Multiple use includes ranching and, according to ranchers, wild horses outcompete their cattle for forage, and therefore adversely affect their herds. Thousands of horses are gathered, removed, and adopted off to the public annually [Bureau of Land Management 2010]. The fate of adopted animals is not tracked. On Assateague, management by the National Park Service (NPS) began primarily in response to an increase in the population from 21 horses in 1965 to 167 horses in 1975. Although the population remained relatively stable for the next nine years [Kirkpatrick 1995], concern over the animals' potential impact on the island's ecology prompted the NPS to quantify their effects on island vegetation. Horses seemed to have little impact on some plant species [Rodgers 1985], but serious impacts on others [Furbish 1980]. In 1986, research on the potential to manage the population with fertility control was initiated, and in 1994 the contraception program was officially begun. Such studies were not conducted on Shackleford Banks, but a more than doubling of the population over ten years from 100 horses in 1985 to ~220 in 1995 [Rubenstein and Dobson, unpublished report], prompted concern over the horses' potential impact on the island's vegetation, and resulted in their eventual management. The contraception program on Shackleford Banks began in the spring of 2000.

Management organizations have limited options to control wild horse populations. Lethal methods are not accepted by the general public and are currently illegal in several states. Gather and removal, also criticized by the public, is commonly conducted by helicopter, all terrain vehicle, "Judas" horse, and/or some combination of these. Horses are herded, often from great distances, into corrals. In some cases, efforts are made to keep harem groups together, in others the groups are split, male from female, foal from dam. In all cases, horses are confined to a relatively small area (often for several days before they are removed and either sold or adopted off) and are put in direct contact with individuals they would not normally encounter in the wild. For a socially complex species like the horse, the experience is highly stressful. Moreover, the removal of the animals permanently affects the genetic makeup of the remaining herd and increases the reproductive efficiency of the horses that remain [Keiper and Houpt 1984, Kirkpatrick and Turner 1991, Cameron et al. 2003].

In the mid-late 80s, research investigating the use of immunoantifertility achieved contraception by raising antibodies against 1) gonadotropin releasing hormone (GnRH) in both males and females, 2) spermatozoa, and 3) ovarian zona pellucida [Turner and Kirkpatrick 1991]. The latter, which did not affect hormone levels in the target animal, was investigated more intensively and the porcine zona pellucida (PZP) vaccine was developed. Females of several species including mice [Jilek and Pavlok 1975], hamsters [Oikawa and Yanagimachi 1975], and rabbits [Sacco 1977] immunized with PZP showed reduced fertility. Studies with wild horses on Assateague Island National Seashore proved particularly promising. PZP was highly effective, reversible, could be remotely delivered, and had only low-level side effects [Kirkpatrick 1990]. These results were especially well received, given the alternatives to wild horse population management.

Effects of PZP in Wild Horses

Behavior

We know very little about the potential for PZP to affect recipient behavior. Powell's study on Assateague Island National Seashore [1999] represents one of the few studies to systematically address this question. He found that recipient and non-recipient mares did not differ in their activity budgets or their spatial relationships relative to the harem stallion. This is in contrast to what has been found with Shackleford Banks horses. Finer-scale measures of activity budgets show an interaction with age such that younger mares (aged 2-3 years) receiving PZP spent less time grazing and more time standing than did non-recipient mares [Rogers 2001]. Although treated mares are often in better physical condition than untreated mares [Kirkpatrick et al. 1996], decreases in the performance of important maintenance behaviors may present problems for the animals' long-term health [Sadleir 1969, Ransom et al. 2010].

In addition, during the non-breeding season, recipient mares on Shackleford exhibit differences in their spatial relationships within the harem. They spend a greater percentage of time in the center of the group and exhibit closer distances to the harem male, on average, than do non-recipient mares (Generalized Linear Model: mare placement, $P = 0.03$; distance to the harem male, $P = 0.007$, see Figure 1, A and B). This effect is significant even when controlling for mare and harem male age, group size, and the presence/absence of recipient and/or non-recipient mares. Moreover, in harems containing both contracepted and uncontracepted females, males are more likely to have contracepted females as their near neighbors (Generalized Linear Model: $P = 0.05$, see Figure 1, C). Perhaps more striking, is the fact that during both the breeding and non-breeding season, Shackleford mares receiving PZP change harem groups more often and visit more groups than do non-recipients [Nunez et al. 2009, Smith 2010]. Such changes in behavior can adversely affect harem group stability [Feist and McCullough 1976, Rubenstein 1981, Goodloe et al. 2000] and therefore, have the potential to affect the overall health and well-being of all harem group members [Kaseda et al. 1995, Linklater et al. 1999]. The addition of new females (especially strangers) often disturbs resident females [Monard and Duncan 1996], resulting in an overall increase in their aggression [Rutberg 1990, Monard and Duncan 1996]. Additionally, repeated changes to harem composition are likely to preclude the establishment of stable female dominance hierarchies, which are necessary for the maintenance of social cohesion among mares and overall group stability [Berger 1977, Houpt and Wolski 1980, Heitor et al. 2006]. The instability caused by switching females may reduce group cohesion even further by adversely affecting the resident females' relationship with the harem male. Finally, because of their tendency to interact with several different groups, the effects of contracepted mares on harem stability level are potentially widespread throughout the population.

Recipient mares on Shackleford also solicit and receive more instances of reproductive behavior than do non-recipients during the non-breeding season [Nunez et al. 2009]. Similar results regarding reproductive behavior have been found in three discrete populations of wild horses in the western United States, with recipient females receiving 54.5% more reproductive behaviors than non-recipients [Ransom et al. 2010]. Moreover, the factors determining the rate of reproductive behavior expressed toward non-recipients (by the harem male) do not affect the rates expressed toward recipient mares [Ransom et al. 2010]. With non-recipient mares, harem males show an increase in the rate of reproductive behavior toward females aged 9-14 years that are more likely to produce a viable foal. This relationship breaks down with recipient females: the harem males' rate of reproductive behavior is both higher and shows no clear pattern among age cohorts or the likelihood of viable offspring production. Such changes in harem male behavior in response to female treatment indicate the potential pervasiveness of PZP's effects in social species.

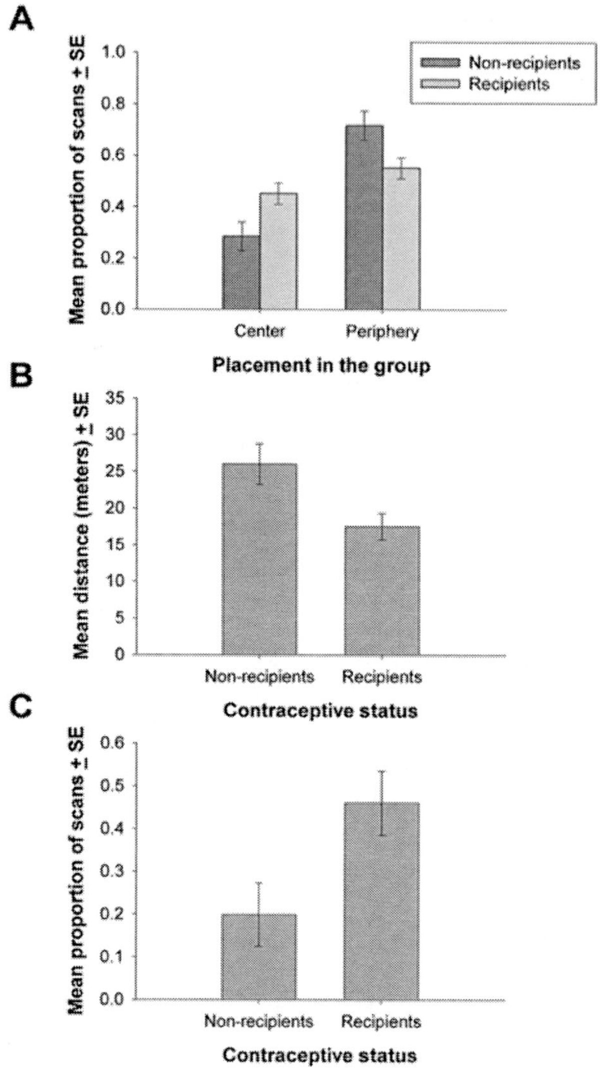

Figure 1. Mares receiving PZP exhibit different spatial relationships with the harem male than do non-recipients. (A) the mean proportion of scans in which mares were located in the center or on the periphery of the group, P=0.03; (B) the mean distance between mares and the harem male, P=0.007; (C) the mean proportion of scans in which mares were the harem male's nearest neighbor, P=0.05 Data were analyzed by Generalized Linear Model. Initial variables in the model included mare and harem male age, group size, and the presence/absence of recipient and/or non-recipient mares. Nonsignificant variables were removed by backwards elimination [see Nunez et al. 2009].

The differences in the results of these studies may be due to a variety of factors including the individual populations' demography and ecology, and the time of study (breeding vs. non-breeding season). Moreover, it should be noted that 18 of the 19 "non-recipients" in the Assateague population had received PZP for the three seasons previous to the study [Powell 1999]. It is not likely that these animals constitute true controls as their behavior may have been affected by the previous contraceptive treatment [Nunez et al. 2009]. The non-recipients on Shackleford Banks and in the three western populations had never received PZP during their lifetime. It is therefore likely that data on these animals more accurately represents the natural behavior of uncontracepted individuals.

In addition, it is likely that the scheduling of PZP administration has contributed strongly to the differences in recipient response. In 1995, after nine years of research, an environmental assessment was prepared for management-level application of PZP to the Assateague herd [Kirkpatrick 1995]. The priorities for treatment followed a hierarchical approach that was based on prior breeding success of the population to ensure that all mares were given an opportunity to reproduce. Females for which there was a high priority for treatment included those that had either a) produced two or more generations of surviving offspring, b) produced more than one surviving offspring, or c) produced one surviving offspring. Low priority females included those that were aged less than four years. Females greater than four years old that had not produced surviving offspring did not receive treatment. In addition, the plan stipulated that only mares that had produced at least three surviving offspring or two generations of offspring would receive more than three consecutive years of treatment. Foals were not to be removed due to the subsequent effects on mare fecundity [Keiper and Houpt 1984, Kirkpatrick and Turner 1991, Cameron et al. 2003]. Finally, it was recognized that this plan was subject to change as the population numbers decreased [Kirkpatrick 1995]. As of 2006, recipient mares were being contracepted upon reaching maturity and were contracepted for three consecutive years. Mares were then allowed to breed normally until one foal was produced, after which time they were contracepted for the rest of their reproductive lives [Zimmerman et al. 2006].

Upon the inception of contraception management on Shackleford Banks in 2000, eight mares were designated as controls based on their age, prior breeding success, and genetic representation in the population. In a given year, ~60% of the foals birthed by these mares were removed [Nunez et al. 2009]. Control mares were not meant to receive the vaccine during their lifetime, but were administered the vaccine in 2009, primarily because of their increased productivity [C. Mason, personal communication], likely due to their offspring's removal [Keiper and Houpt 1984, Kirkpatrick and Turner 1991]. Regardless of their productivity, mares designated as recipients receive contraception upon reaching two years of age and are administered an average of 3.39 ± 0.17 consecutive years of contraception, after which, any foals born are likely to be removed. This difference in PZP administration may play a key role in the responses of Shackleford mares, as the ability to conceive after dispersal is likely critical, not only to establishing lasting fidelity to a new harem male, but also to normal, physiological functioning [Turner and Kirkpatrick 1991].

Physiology

Prior research has shown little to no effect of PZP on wild horse physiology [Kirkpatrick et al. 1996, Kirkpatrick et al. 1997, Powell and Monfort 2001]. For example, on Assateague, PZP has had no effect on the duration of individual estrous cycles [Powell and Monfort 2001], and as many as seven consecutive years of treatment resulted in no debilitating side effects. In addition, the contraceptive effects of PZP proved reversible after up to four consecutive years of treatment, and even long-term treatment (5-7 years) caused only minor ovulation failure and depressed urinary oestrogen concentrations [Kirkpatrick et al. 1996].

Finally, PZP proved safe for pregnant females, and did not adversely affect the survivorship or subsequent fertility of offspring born to treated individuals [Kirkpatrick and Turner 2002]. However, behavioral research on Shackleford Banks horses [Nunez et al. 2009], white-tailed deer [McShea et al. 1997], and free-ranging elk [Heilmann et al. 1998] suggests that PZP may affect the reproductive physiology of recipient animals. In each of these studies, females treated with PZP extended reproductive behaviors into what is typically the non-breeding season. These results are indicative of an extension of ovulatory cycling into the post-breeding season, when most females are normally anovulatory.

Gestation in wild horses lasts approximately 11-12 months [Asa 2002, see above]. It is therefore possible to reliably estimate conception date from foaling date. Since the contraception program began on Shackleford Banks, foaling has occurred later and over a significantly broader range than it had before the program [Nunez et al. in review]. This effect is still significant even when controlling for mare age and physical condition, temperature, rainfall, and the presence/absence of other management practices [Nunez et al. in review, see Figure 2, A]. Although the effect is more prevalent in recipient mares (particularly those receiving several consecutive applications), this change is also exhibited by non-recipients. A similar study conducted on Assateague did not find such differences [Kirkpatrick and Turner 2003, see Figure 2, B]. Management regimes at the two sites differ and may contribute to these divergent patterns (see above). Additionally, the Assateague study did not discriminate between foaling patterns pre- and post- the contraception program, which may have masked some important effects.

By preventing pregnancy without altering hormonal cycling, effective PZP treatment, by default, increases the average number of ovulatory cycles a mare undergoes each breeding season [Powell 1999]. Typically a mare will conceive during the breeding season after which time she ceases to cycle. This is not the case for a successfully contracepted mare; she will continue to cycle throughout the breeding season [Asa 2002]. Therefore, PZP treatment likely extends the reproductive behavior of both treated mares and their harem males throughout the whole of the breeding season [Powell 1999]. Changes in foaling date indicate a more dramatic change in the schedule of ovulatory cycling, whereby recipient mares extend reproductive physiology into what is typically the non-breeding season.

Such changes have the potential to disrupt behaviors important to sufficient nutrient acquisition by all group members and may, therefore, have both social and demographic consequences. In temperate environments, food availability is lowest during the fall and winter months [Sadleir 1969]. During these seasons, free-ranging horses increase the time spent foraging (at the expense of other activities) which likely increases consumption and lowers their overall energetic costs [Duncan 1985, Morel et al. 2006]. Prior to contraception management, Shackleford horses increased inter-individual spacing during this time of year which likely reduced competition [personal observation].

The extension of cycling into the post-breeding season by mares increases male attentiveness [Nunez et al. 2009] which may limit female movement, decrease group spread, and reduce foraging efficiency [Tyler 1972, Feist and McCullough 1976]. In addition, this more active defense of females during the fall/winter months likely requires higher energy expenditure by males during harsh environmental conditions [Stevens 1990] and could lead to a reduction in their body condition [Ransom et al. 2010]. Offspring conceived (and born) during what is normally the post-breeding season may also be subject to decreased resources since low quality forage likely affects mares' ability to produce sufficient milk [Sadleir 1969], but see Kirkpatrick and Turner [2003].

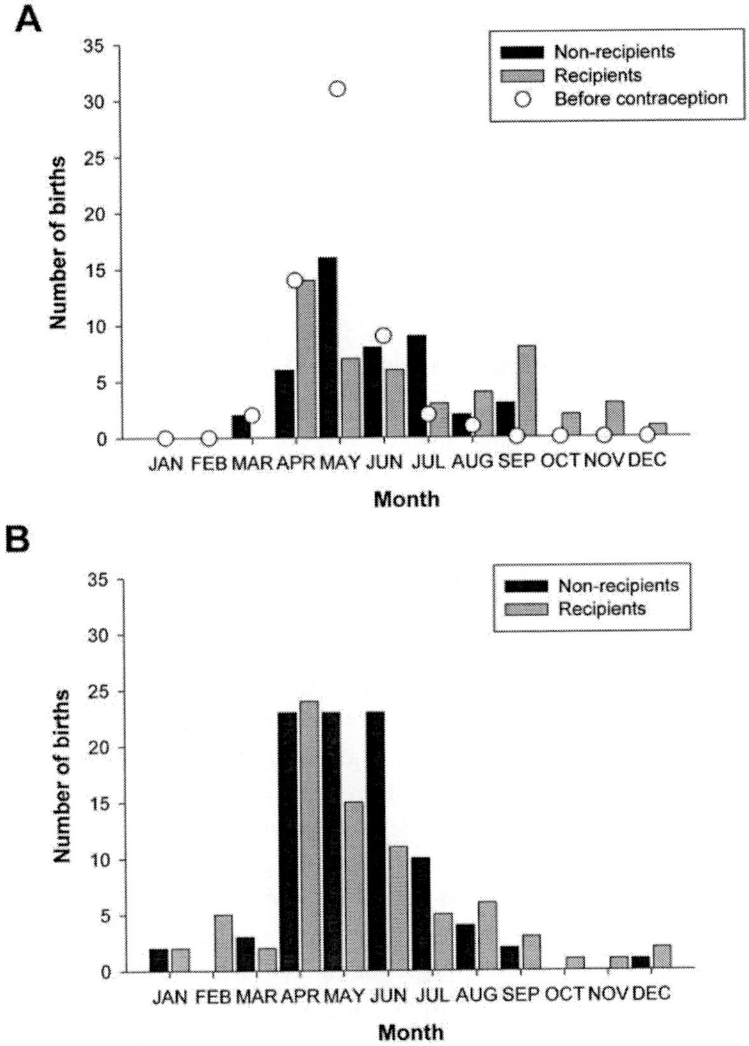

Figure 2. Distribution of foaling dates for recipient and non-recipient mares on (A) Shackleford Banks, NC and (B) Assateague Island National Seashore, MD. Assateague data redrawn from Kirkpatrick and Turner, 2003.

The increased reproductive behavior between recipient mares and their stallions may have the additional effect of entraining non-recipients to continue reproductive behaviors and cycling into the non-breeding season. Such entrainment would help explain why even non-recipients have shown shifts in reproductive cycling after PZP management on Shackleford Banks [Nunez et al. in review]. Reproductively active conspecifics are commonly used to induce receptivity in several domestic species including horses [Davies Morel 2003], pigs [Morris 2001], and cows [Hornbuckle et al. 1995]. In the wild, courtship signals from conspecifics advance gonadal cycles or maturation in several taxa, including mammals [Marsden and Bronson 1964, Whitten et al. 1968, McComb 1987], birds [Ball and Balthazart

2002], amphibians [Lea et al. 2001], and reptiles [Lindzey and Crews 1988]. The potential for PZP treatment to thusly affect the reproductive physiology of multiple individuals within a population, regardless of treatment status, is an important consideration.

CONCLUSION

When the alternatives are considered, PZP is currently managers' most humane and effective option for population control. PZP has been used to control the Assategue Island population for 16 years, and recipient mares show no behavioral and only minor physiological responses [Kirkpatrick et al. 1996, Kirkpatrick et al. 1997, Powell 1999, Powell and Monfort 2001, Kirkpatrick and Turner 2002, Turner et al. 2002]. However, horses on Shackleford Banks [Nunez et al. 2009] and in the western United States [Ransom et al. 2010] alter social and reproductive behaviors in response to PZP. These differences indicate that not all populations will respond similarly to PZP treatment. Differences in habitat, resource availability, and demography among conspecific populations will undoubtedly affect their physiological and behavioral responses to PZP contraception, and need to be considered. Careful study of the animals' physiology, behavior, and demography is necessary prior to and during treatment to ensure that potential effects of PZP can be assessed accurately and that, within managerial constraints, PZP effects are ameliorated as much as is possible.

A thorough understanding of a species' biology and behavioral ecology is imperative in the planning stages of contraception scheduling. Seemingly small differences in the management regime could lead to very different and potentially detrimental outcomes. If feral horse herds are to be maintained in as natural a state as possible, we recommend that sub-adult, dispersing females be allowed to have (and keep) at least one foal before receiving contraception [Kirkpatrick 1995], and that mares should not receive PZP for several consecutive seasons. Such management regimes would of course necessitate a higher minimum population level and may involve additional costs. However, given the expense of removing foals and the efforts required to find safe homes for them, allowing mares to keep their offspring may offset these costs. Moreover, given the potential benefit to the health and well-being of both target and non-target animals, these strategies are worth serious consideration. For a social species like the horse, they could bridge the gap between managing population size and maintaining behaviorally functional populations.

REFERENCES

Asa, C. S. (2002). Equid reproductive biology. Pages 113-117 in: *IUCN/SSC Equid Action Plan*, P. D. Moehlman, editor. Gland, Switzerland.

Asa, C. S., Goldfoot, D. A. & Ginther, O. J. (1979). Sociosexual behavior and the ovulatory cycle of ponies (*Equus caballus*) observed in harem groups. *Hormones and Behavior, 13*, 49-65.

Asa, C. S., Goldfoot, D. A., Garcia, M. C. & Ginther, O. J. (1980). Sexual behavior in ovariectomized and seasonally anovulatory pony mares (*Equus Caballus*). *Hormones and Behavior*, *14*, 46-54.

Ball, G. F. & Balthazart, J. (2002). Neuroendocrine mechanisms regulating reproductive cycles and reproductive behavior in birds. Pages 649-798 In: D. W. Pfaff, A. P. Arnold, A. M. Etgen, S. E. Fahrbach, & R. T. Rubin, editors. *Hormones, brain and behavior.* Academic Press,San Diego.

Berger, J. (1977). Organizational systems and dominance in feral horses in Grand Canyon. *Behavioral Ecology and Sociobiology*, *2*, 131-146.

Bureau of Land Management. (2010). Bureau of Land Management Wild Horse and Burro Statistics http://www.wildhorseandburro.blm.gov/statistics/1996stat.htm.

Cameron, E. Z., Linklater, W. L., Stafford, K. J. & Minot, E. O. (2003). Social grouping and maternal behaviour in feral horses (*Equus caballus*): the influence of males on maternal protectiveness. *Behavioral Ecology and Sociobiology*, *53*, 92-101.

Churcher, C. S. (1993). *Equus grevyi*. Pages 1-9 in: *Mammalian Species*.

Crowell-Davis, S. L. (2007). Sexual behavior of mares. *Hormones and Behavior*, *52*, 12-17.

Davies Morel, M. C. G. (2003). *Equine reproductive physiology, breeding, and stud management,* 2nd edition. CABI Publishing, Oxon, UK.

Duncan, P. (1985). Time-budgets of Camargue horses; environmental influences. *Behaviour*, *92*, 188-208.

Feist, J. D. & McCullough, D. R. (1976). Behavior patterns and communication in feral horses. *Zeitschrift für Tierpsychologie*, *41*, 337-371.

Furbish, C. E. (1980). *Factors affecting the distribution of Distichlis spicata in the Spartina altiniflora saltmarshes of Assateague Island, Maryland.* University of Maryland Eastern Shore, Princess Anne.

Ginther, O. J. (1979). *Reproductive Biology of the Mare: Basic and Applied Aspects.* Equiservices, Cross Plains, WI.

(1992). *Reproductive Biology of the Mare: Basic and Applied Aspects,* 2nd ed. Equiservices, Cross Plains, WI.

Goodloe, R. B., Warren, R. J., Osborn, D. A. & Hall, C. (2000). Population characteristics of feral horses on Cumberland Island, Georgia and their management implications. *Journal of Wildlife Management, 64*, 114-121.

Grubb, P. (1981). *Equus burchelli*. Pages 1-9 in: *Mammalian Species*.

Heilmann, T. J., Garrott, R. A., Cadwell, L. L. & Tiller, B. L. (1998). Behavioral response of free-ranging elk treated with an immunocontraceptive vaccine. *Journal of Wildlife Management, 62*, 243-250.

Heitor, F., Oom, M. D. & Vicente, L. (2006). Social relationships in a herd of Sorraia horses Part 1. Correlates of social dominance and contexts of aggression. *Behavioural Processes*, *73,* 170-177.

Hornbuckle, T., Ott, R. S., Ohl, M. W., Zinn, G. M., Weston, P. G. & Hixon, J. E. (1995.) Effects of bull exposure on the cyclicity activity of beef cows. *Theriogenology*, *43*, 411-418.

Houpt, K. A. & Wolski, T. R. (1980). Stability of equine hierarchies and the prevention of dominance related aggression. *Equine Veterinary Journal*, *12*, 15-18.

Jilek, F. & Pavlok, A. (1975). Antibodies against mouse ovaries and their effect on fertilization invitro and invivo in the mouse. *Journal of Reproduction and Fertility*, *42*, 377-380.

Kaseda, Y., Khalil, A. M. & Ogawa, H. (1995). Harem stability and reproductive success of Misaki feral mares. *Equine Veterinary Journal*, *27*, 368-372.

Keiper, R. & Houpt, K. (1984). Reproduction in feral horses: an 8 year study. *American Journal of Veterinary Research*, *45*, 991-995.

Kirkpatrick, J. (1995). *Management of wild horses by fertility control: the Assageague experience*. United States Department of the Interior, National Park Service.

Kirkpatrick, J. & Turner, J. W. (1991). Compensatory reproduction among feral horses. *Journal of Wildlife Management*, *55*, 649-652.

Kirkpatrick, J. F., Liu, I. M. K. & Turner J. W., Jr. (1990). Remotely-delivered immunocontraception in feral horses. *Wildlife Society Bulletin*, *18*, 326-330.

Kirkpatrick, J. F., Liu, I. M. K., Turner, J. W., Naugle, R. & Keiper, R. (1992). Long-term effects of porcine zonae pellucidae immunocontraception on ovarian function in feral horses (*Equus Caballus*). *Journal of Reproduction and Fertility*, *94*, 437-444.

Kirkpatrick, J. F. & Turner, A. (2002). Reversibility of action and safety during pregnancy of immunization against porcine zona pellucida in wild mares (*Equus caballus*). *Reproduction*:197-202.

(2003). Absence of effects from immunocontraception on seasonal birth patterns and foal survival among barrier island wild horses. *Journal of Applied Animal Welfare Science, 6*, 301-308.

Kirkpatrick, J. F., Turner, J. W., Liu, I. K. M. & FayrerHosken, R. (1996). Applications of pig zona pellucida immunocontraception to wildlife fertility control. *Journal of Reproduction and Fertility*, 183-189.

Kirkpatrick, J. F., Turner, J. W., Liu, I. K. M., Fayrer-Hosken, R. & Rutberg, A. T. (1997). Case studies in wildlife immunocontraception: Wild and feral equids and white-tailed deer. *Reproduction Fertility and Development*, *9*, 105-110.

Klingel, H. (1975). Social organization and reproduction in equids. *Journal of Reproduction and Fertility*, 7-11.

Lea, J., Dyson, M. & Halliday, T. (2001). Calling by male midwife toads stimulates females to maintain reproductive condition. *Animal Behaviour*, *61*, 373-377.

Lindzey, J. & Crews, D. (1988). Psychobiology of sexual behavior in a whiptail lizard, *Cnemidophorus inornatus*. *Hormones and Behavior*, *22*, 279-293.

Linklater, W. L., Cameron, E. Z., Minot, E. O. & Stafford, K. J. (1999). Stallion harassment and the mating system of horses. *Animal Behaviour*, *58*, 295-306.

Linklater, W. L., Cameron, E. Z., Stafford, K. J. & Veltman, C. J. (2000). Social and spatial structure and range use by Kaimanawa wild horses (*Equus caballus*: Equidae). *New Zealand Journal of Ecology*, *24*, 139-152.

Marsden, H. M. & Bronson, F. H. (1964). Estrous synchrony in mice-alteration by exposure to male urine. *Science*, *144*, 1469.

Mason, C. (2009). personal communication.

McComb, K. (1987). Roaring by red deer stags advances the date of estrus in hinds. *Nature*, *330*, 648-649.

McShea, W. J., Monfort, S. L., Hakim, S., Kirkpatrick, J., Liu, I., Turner, J. W., Chassy, L. & Munson, L. (1997). The effect of immunocontraception on the behavior and reproduction of white-tailed deer. *Journal of Wildlife Management*, *61*, 560-569.

Monard, A. M. & Duncan, P. (1996). Consequences of natal dispersal in female horses. *Animal Behaviour*, *52*, 565-579.

Morel, M., McBride, S. D., Chiam, R. S., McKay, A. & Ely, E. (2006). Seasonal variations in physiological and behavioural parameters in a bachelor group of stallion ponies (*Equus caballus*). *Animal Science*, *82*, 581-590.

Morris, J. (2001). Vasectomy in the pig. *In Practice*, *23*, 606-610.

Nunez, C. M. V., Adelman, J. S., Mason, C. & Rubenstein, D. I. (2009). Immunocontraception decreases group fidelity in a feral horse population during the non-breeding season. *Applied Animal Behaviour Science*, *117*, 74-83.

Nunez, C. M. V., Adelman, J. S. & Rubenstein, D. I. (2010). Immunocontraception in wild horses (*Equus caballus*) extends ovulatory cycling beyond the usual foaling season. In review.

Oikawa, T. & Yanagimachi, R. (1975). Block of hamster fertilization by antiovary antibody. *Journal of Reproduction and Fertility*, *45*, 487-494.

Penzhorn, B. L. (1988). *Equus zebra*. Pages 1-7 in: *Mammalian Species*.

Powell, D. M. (1999). Preliminary evaluation of porcine zona pellucida (PZP) immunocontraception for behavioral effects in feral horses (*Equus caballus*). *Journal of Applied Animal Welfare Science*, *2*, 321-335.

Powell, D. M. & Monfort, S. L. (2001). Assessment: Effects of porcine zona pellucida immunocontraception on estrous cyclicity in feral horses. *Journal of Applied Animal Welfare Science, 4*, 271-284.

Ransom, J. I., Cade, B. S. & Hobbs, N. T. (2010). Influences of immunocontraception on time budgets, social behavior, and body condition in feral horses. *Applied Animal Behaviour Science*, *124*, 51-60.

Rodgers, R. B. (1985). *Assateague Island National Seashore Feral Pony Management Plan.*

Rogers, J. E. (2001). *The effects of immunocontraception on the social behavior of female feral horses* (*Equus caballus*). Princeton University, Princeton.

Rubenstein, D. I. (1981). Behavioral ecology of island feral horses. *Equine Veterinary Journal, 13*, 27-34.

(1986). Ecology and sociality in horses and zebras. Pages 282-302 in: *Ecological Aspects of Social Evolution, Birds and Mammals*. D. I. Rubenstein, Wrangham, R.W. editor. Princeton University Press, Princeton.

Rubenstein, D. I. & Dobson, A. P. (2000). *Demographic and genetic dynamics of island feral horses: implications for fertility control*. Princeton University.

Rutberg, A. T. (1990). Intergroup transfer in Assateague pony mares. *Animal Behaviour*, *40*, 945-952.

Sacco, A. G. (1977). Antigenic cross-reactivity between human and pig zona pellucida. *Biology of Reproduction*, *16*, 164-173.

Sadleir, R. M. F. S. (1969). *The Ecology of Reproduction in Wild and Domestic Mammals.* Methuen & Co. Ltd., London.

Smith, J. (2010). *The inoculation of porcine zona pellucida and its behavioral and physiological effects on the feral mares of Shackleford Banks*. Princeton University, Princeton.

Stevens, E. F. (1990). Instability of harems of feral horses in relation to season and presence of subordinate stallions. *Behaviour*, *112*, 149-161.

Turner, J. W. & Kirkpatrick, J. F. (1991). New developments in feral horse contraception and their potential application to wildlife. *Wildlife Society Bulletin*, *19*, 350-359.

Turner, J. W., Liu, I. K. M., Flanagan, D. R., Bynum, K. S. & Rutberg, A. T. (2002). Porcine zona pellucida (PZP) immunocontraception of wild horses (*Equus caballus*) in Nevada: a 10 year study. *Reproduction*, 177-186.

Tyler, S. J. (1972). The behaviour and social organization of the New Forest ponies. *Animal Behaviour Monographs*, *5*, 85-196.

Weinstock, J., Willerslev, E., Sher, A., Tong, W., Ho, S.Y.W., Rubenstein, D., Storer, J., Burns, J., Martin, L., Bravi, C., Prieto, A., Froese, D., Scott, E., Xulong, L. & Cooper. A. (2005). Evolution, systematics, and phylogeography of pleistocence horses in the new world: A molecular perspective. *Plos Biology*, *3*, 1373-1379.

Whitten, W. K., Bronson, F. H. & Greenstein, J. A. (1968). Estrus-inducing pheromone of male mice - transport by movement of air. *Science*, *161*, 584-585.

Zimmerman, C., Sturm, M., Ballou, J. & Traylor-Holzer, K. (2006). *Horses of Assateague Island population and habitat viability assessment: final report.*

In: Horses: Biology, Domestication, and Human Interactions
Editor: James E. Leffhalm
ISBN: 978-1-61761-303-6

Chapter 5

PATHOGEN INACTIVATION DURING HORSE MANURE COMPOSTING

L. Hébert, S. Petry and C. Laugier
Anses, Laboratoire de pathologie équine de Dozulé
14430 Goustranville, France.

ABSTRACT

Over the past few decades, composting has become one of the most common technological alternatives for horse manure valorization. However, such practices can lead to increased contamination of the farm environment to a greater extent and constitute a risk of infection for both humans and animals, especially if compost is spread in the fields where horses graze. After reviewing bacterial, viral and parasitic pathogens that can be found in horse manure, we shall describe processes that inactivate pathogens during composting, time-temperature regulations and guidelines applicable to horse manure composting operations. We shall then describe factors that could influence the composting processes and different ways that pathogens can be spread in the environment, before evaluating the risk that they represent for horse and human health. Finally, we shall discuss the relevance of international compost quality standards from a practical point of view.

INTRODUCTION

Horse manure is mainly composed of feces and urine, often mixed with bedding material and feed. Large variations in the amount and characteristics of horse manure production can be observed depending on many different factors such as horse type (pony, mature horse, lactating mare, etc.), body weight, feed, exercise levels, bedding material choice and waste collection practices ...We can generally estimate that an average 455-kg horse produces about 14 kg of feces and 11 l of urine, which add up to 23 kg/day, or 9.5 m^3/yr. When bedding material is added, this volume can easily exceed 20 m^3/yr per horse [1]. As a potential source of bacteria and other pathogens, horse manure represents a major health concern. Due to the risk of runoff and leachate from manure polluting watercourses and domestic water supplies, horse manure must be managed properly to reduce risks to both human health and the environment.

Composting is a managed biological process that converts organic material into a dark, crumbly, earthy-smelling end product similar to potting soil. Manure is a nutrient-rich environment favorable to microorganisms whose metabolic activity decomposes and digests the organic material in the presence of oxygen. Heat generated from microorganisms' metabolic activity plays a major role in the inactivation of pathogens [2]. Microorganisms responsible for composting need food, water and air to survive. To ensure successful composting and a good final product, the horse manure's chemical and physical characteristics therefore need to be strictly controlled. Manure piles generally have different percentages of feces, urine, bedding and feed, thus influencing the composting process. Of all the factors influencing composting, the most important are moisture levels, carbon content, oxygen availability and particle size [3]. As an example, a low moisture content slows the composting process whereas a high moisture content reduces open pore space and limits aeration (oxygen). A high carbon content slows down microbial degradation while low carbon content ratios result in the release of nitrogen as ammonia, which is a source of unpleasant odors [3]. Theoretically, a well-made compost stabilizes the organic matter, decreases the volume of waste by 50 %, reduces odor, kills parasite eggs and larvae, destroys weed seeds and transforms a potential liability into a marketable resource allowing it to be spread on pastures. Finished compost can be used to improve soil quality or serve as a mulch or growing media for plant nurseries, gardeners, mushroom growers and worm farms. However, if not done properly, composting can increase contamination of the farm environment and may constitute a risk of infection for both humans and animals, especially if compost is spread in the fields used by grazing horses.

Pathogens Potentially Encountered in Horse Manure

It is not feasible to mention all the pathogens found in horse manure because of the huge number and variety of organisms propagating in the gut and related tissues as well as glands. Moreover, when considering the problem of pathogen transmission through manure, it is important to recognize that manure, which consists of animal excreta (feces and urine), bedding and dilution water, also contains secretions from the nose, throat, blood, vagina, mammary gland, skin, and placenta. Microbes in these secretions, as well as those in the excreta, can potentially accumulate on the barn floor. We report here some pathogens of concern likely to be found in horse manure, along with their potential health effects (Table 1).

Bacteria

Bacteria are of particular interest when considering horse manure composting since most can survive and/or multiply in the environment without a host and up to several years at a time [4]. It should be noted that numerous pathogenic bacteria encountered in horse manure compost can cause zoonotic infections and can be transmitted from animals to humans directly or through food and water supplies. They should thus be the object of particular interest. Among bacteria potentially encountered in horse manure, it appears that *Salmonella spp.*, *Listeria monocytogenes* and *Escherichia coli* (especially O157, H7) are currently a growing public health concern because they have multiple hosts and their prevalence of infection has increased over the past 20 years [2]. Furthermore, bacteria such as *Bacillus* and *Clostridium* produce resistant endospores that can survive for extended periods under numerous environmental conditions and are capable of surviving temperature conditions that inactivate the majority of focally-excreted pathogens [5]. It is therefore unlikely that bacterial endospores would be destroyed by high temperatures and would doubtless survive in compost despite application of current time–temperature regulations. To take an example, *Bacillus cereus* has been isolated after a composting process of 7 days at 60 to 65 °C (although it was not detected once the temperature reached 70 °C)[6].

Viruses

Viruses, which are obligate intracellular parasites, often have a limited host range. They are unable to replicate outside of their host, so their numbers never increase once they are released into the environment. The most commonly diagnosed cause of foal diarrhea in intensive breeding areas worldwide is rotavirus Group A: it has been reported that 15–50% of diarrheal foals shed rotaviruses [7]. These viruses can remain infectious for a long time in the environment and as an infective dose is very small, the survival of as little as one plaque-forming unit should be of concern. Generally speaking, intestinal viruses from farm animals do not generally infect humans thus; viruses are considered to be host–specific and farm animals are not considered a source of infection for humans [2].

Table 1. Some pathogens of interest found in horse manure, associated diseases and zoonotic character

Group	Organism	Associated diseases	Zoonotic	Reference
Bacteria	*Aeromonas spp.*	Diarrhea	O	[26]
	Campylobacter spp.	Diarrhea	O	[27]
	Clostridium difficile	Colitis	O	[28]
	Clostridium tetanii	Tetanus	O	[29]
	Clostridium perfringens	Gangrene, gastroenteritis	O	[28]
	Clostridium botulinum	Botulism	O	[30]
	Escherichia coli and other coliforms	Gastroenteritis, diarrhea	O	[31]
	Listeria monocytogenes	Listeriosis	O	[32]
	Rhodococcus equi	Rhodococcosis	O	[33]
	Salmonella spp.	Salmonellosis	O	[28]
	Streptococcus spp.	Streptococcosis	O	[34]
	Yersinia enterocolitica	Yersiniosis	O	[27]
Parasites	*Cryptosporidium parvum*	Cryptosporidiosis	O	[10]
	Cyathostomins	Cyathostominosis	N	[35]
	Fasciola hepatica	Fasciolosis	O	[36]
	Giardia intestinalis	Giardiasis	O	[10]
	Oxyuris equi	Oxyuriasis	N	[37]
	Parascaris equorum	Parascariasis	N	[38]
	Strongyloides westeri	Strongyloidiasis	N	[27]
	Strongylus spp.	Strongylosis	N	[36]
	Anoplocephala perfoliata	Taeniasis	N	[39]
Virus	Group A Rotaviruses	Diarrhea	N	[28]

Parasites

Internal parasites are of a great concern for horse owners as they are a primary cause of colic and can cause or contribute to many respiratory, digestive, and performance problems. Unless controlled, they are likely to increase and cause the horse severe injury or death. From a practical standpoint, the internal parasites excreted in horse feces representing the greatest health risks are strongyles, ascarids, tapeworms and bots. However, due to their ability to escape from manure by burrowing into the ground once excreted in feces, bots (*Gasterophilus intestinalis*) are not really an issue in horse manure composting [8].

Horses affected the most by parasites are young sucklings, weanlings and yearlings. In general, ascarid problems are restricted to young horses because, in most cases, resistance or immunity is built up by the time a horse is 2 to 3 years old. On the other hand, strongyles and bots affect horses of all ages. Even so, younger horses are much more severely affected than older ones.

Protozoans such as G*iardia spp.* and *Cryptosporidium parvum* can also be found in horse manure [9]. In healthy, mature animals (or humans), the infections caused by both *Giardia spp.* and *C. parvum* are usually self-limiting and, although they cause significant discomfort, do not pose serious long-term health risks [10]. However, as G*iardia spp.* and *C.parvum* have an extremely low infective dose (a single protozoan cyst has been observed to cause infection)[4], the survival of even a single cyst is cause for concern. Moreover, cysts or oocysts of these pathogens can resist many environmental pressures, enabling them to remain viable in the environment for at least a year [11]. Of horse parasites, only G*iardia spp.* and *C. parvum* (both zoonotic pathogens) represent a potential risk for humans as they cause severe diarrhea [8].

Pathogen Inactivation during Composting

One of the main goals of a composting process is to reduce pathogenic organisms. It is generally accepted that pathogens must be reduced below detectable levels in the finished product to minimize risks for public health and the environment. The primary process criteria used worldwide for ensuring the microbiological safety of composts have been narrowly defined time–temperature conditions [4]. During composting, pathogens are reduced to some degree by a combination of several processes, including competition between indigenous microorganisms and pathogens, antagonistic relationships between organisms, the action of antibiotics produced by certain fungi and actinomycetes, natural die-off in the compost environment, production of toxic byproducts such as gaseous ammonia, nutrient depletion, and thermal destruction [5]. Because the degree to which pathogens are inactivated by these processes has not been determined, except for thermal destruction, an elevated temperature is considered to be the most useful method for sanitizing compost [5]. High temperatures have profound effects on the structural and physiological properties of pathogens, including membranes, RNA, DNA, ribosomes, protein and enzymes [12]. Thus, it is obvious that no single alteration within an organism is responsible for the lethal effect of high temperature. However, for viruses, it appears that thermal inactivation occurs as a result of damage to the viral structure: specifically, surface proteins are drastically modified when exposed to high temperatures. Once modified, the viruses are no longer infective because they are unable to bind to a host cell [13].

Legislation

In order to reduce the risk of transferring pathogens from horse manure to the environment, several international standards are available to help meet pathogen limits through composting. Table 2 shows some of these guidelines, applicable to composting operations in different countries. As an example, The United States Environmental Protection Agency (US-EPA) publication [4] on controlling pathogens in biosolids describes two classes (A and B) of a finished product. Class A products are designed as treated biosolid products having no detectable pathogens and not restricted in use. To be considered a Class A compost, composting operations must last for 15 days or more at a temperature above 55°C. Windrow piles must be turned five times during the same period. Treated biosolids are classified Class B if pathogens are detectable but have been reduced to levels that would not pose a significant risk for public health or the environment as long as exposure to the product is minimized [4]. This requires the compost pile to be maintained at a minimum of 40 °C for at least five days and, during this five-day period, temperature must rise above 55°C for at least 4 h. For these products, site restrictions apply to prevent crop harvesting, animal grazing and public access for a specified amount of time after application or disposal. Moreover, US-EPA emphasizes that the above time–temperature criteria should apply to every particle of material composted; it is therefore important that temperature monitoring be representative of all the areas in a compost heap and that the temperature profile at each and every monitoring point comply with time–temperature requirements. However, no guidance is given in terms of exact numbers of locations to monitor, specific monitoring locations (aside from expected cool zones in static piles and in-vessel systems) or frequency of monitoring.

Table 2. Time-temperature requirements of different countries for sanitizing compost products. Adapted from [40]

Country	Compost Method	Time/Temperature requirement
US-EPA		> 55 °C for 15 days with at least 5 turnovers
Australia	All methods	> 55 °C for 3 days
Germany	Open Windrow	> 55 °C for 2 weeks or > 65 °C for 1 week
Austria	All composts	> 60 °C for 6 days, or > 65 °C for 3 days, or >65 °C for 2 x 3 days
Switzerland		> 55 °C for 3 weeks, or > 60 °C for 1 week
Denmark	All composts	> 55 °C for 2 weeks

FACTORS INFLUENCING PATHOGEN INACTIVATION DURING HORSE MANURE COMPOSTING

Even if time-temperature requirements have obviously been satisfied within the compost pile, different factors can lead to the presence of pathogens in compost. Of these factors, we will detail temperature heterogeneity during composting, differences between *in vitro* and *in vivo* analysis and regrowth.

Variation between in vitro and in vivo Analysis

The existence of heterogeneous temperatures over time within bulk compost indicates that laboratory-scale studies with a uniform temperature would suggest shorter inactivation times than may be necessary in full-scale systems [14] and thus may be insufficient to accurately predict indicator and (or) pathogen survival times in full-scale composting facilities. For example, the examination of *Escherichia coli*, coliphage, *Salmonella spp.*, and viable *Ascaris lumbricoides* ova in both pilot- and full-scale systems show that in the pilot-scale reactors, concentrations of all the pathogens of interest dropped to below detection limits within eight days or less, whereas during full-scale windrow composting, it took significantly longer to reduce organisms than during the pilot-scale experiments: over two weeks were needed before *Salmonella spp.* was reduced to non-detectable levels, coliphage reduction took at least three weeks, *E. coli* over five weeks and *Ascaris* ova over six weeks [14]. Comparable results were obtained by other studies [15, 16].

Temperature Heterogeneity during Composting

The most commonly cited explanation for pathogen survival during composting is that temperature may not be uniform throughout the entire mass of compost. In other words, temperature monitoring and (or) reporting may incorrectly give the impression that time–temperature criteria have been met throughout the entire composting mass, whereas some zones may never have reached the recommended temperatures. The outer layers, or boundaries, of the pile are the coolest areas, especially the bottom. This fact can be qualitatively observed during winter-time composting, when layers of ice and (or) snow can be observed on the outer surfaces of piles, while the pile interiors experience thermophilic temperatures [5]. Logically, the lower temperatures in outer parts of the pile affect pathogen survival. To take an example, *salmonellae* survival was observed in the bottom corner of an aerated static pile, which was found to be at a lower temperature than the rest of the compost heap [17]. Moreover, since temperatures in compost fluctuate over time, in the short- and long-term [18], two temperature measurements of 55 °C or more taken days apart do not necessarily imply that time–temperature criteria have been met. Between measurements, the temperature could have decreased, as fluctuations of up to 15 °C per day are not uncommon in windrows [19]. Thus, it should be emphasized that the ability to ensure the whole compost pile complies with required time–temperature conditions is as important as the time–temperature conditions themselves.

Regrowth

Regrowth indicates that either some organisms survived composting or that recontamination has occurred. Low-temperature zones within compost are a concern not only because pathogens may not be sufficiently reduced, but because conditions may even be favorable for bacterial pathogen growth. Some bacteria can rapidly proliferate under favorable conditions, especially when competitive bacterial populations have been mainly eliminated through composting. However, this phenomenon cannot explain the presence of viruses, protozoa, or helminths in finished composts since, once their populations are reduced by composting, they cannot regrow outside their specific host organism(s). Moreover, an increase in pathogenic bacteria populations in compost is closely associated with the availability of digestible carbon and low competitive microbiological diversity [20]. Therefore, the regrowth ability of indicator organisms indicates compost immaturity at the end of the sanitation phase.

Pathogen Spreading during Composting

Pathogens in horse manure pose a disease risk only if there are routes by which they are brought into contact with humans or animals. Pathogen spreading during composting is very important since pathogens in soil and plant surfaces can survive for long periods of time. Even if it seems difficult to establish a maximum survival time of pathogens in soil, due to the various physico-chemical conditions encountered in the environment, it has been reported for example that helminth ova can survive up to seven years in soil [4]. Of the various ways that pathogens can be spread during manure composting, we can name vector attraction, aerosolization, fomites and water runoff.

Vector Attraction

US-EPA describes vectors as any living organism capable of transmitting a pathogen from one organism to another either mechanically by simply transporting the pathogen on an animal's body or biologically by playing a specific role in the life cycle of the pathogen (*e.g.* replication in the other animal's body)[4]. Vectors for horse manure pathogens mainly include insects, rodents and birds. However, the main limitation for the study of this parameter to date is the lack of suitable methods for measuring vector attraction directly. It appears that vector attraction can be significantly reduced by employing physical barriers between vectors and compost.

Water Runoff

Water runoff is an important potential way of transporting pathogens from manure to water bodies. It is vital to limit runoff to avoid a composting area contributing to ground or surface water pollution. As an example, a covered storage area will have much less leachate than one exposed to precipitation, and a concrete pad with sidewalls would be necessary to contain leachate from very large, uncovered piles. This is of special concern for manure from animals up to six months old, which represents a potential source of *Giardia spp.* and *C. parvum* able to cause infection in humans and horses [4].

Aerosolization

The aerosolization of protozoa and helminths is unlikely, but bacteria and viruses may become airborne and disperse from a source [4]. This phenomenon depends on local meteorological and topographical conditions, and indicates that microbes that become airborne (via aerosols, dust, etc.) can be inhaled when turning over compost, for example, or during strong winds, plowing or cultivating the soil once spread. It is suggested that persons at risk avoid the compost area during turning operations. They should also wear a dust mask when performing operations that generate dust. Such operations should also be carried out in the absence of horses.

Fomites

Fomites are inanimate objects that can harbor a pathogen and serve as its transmission agent. To take the example of horse manure composting, fomites may be the equipment, boots or clothing of compost operators. There are so many objects that could be fomites, it is difficult to give specific instructions for every case, but most of the time, the establishment of good working protocols including a program of terminal cleaning and disinfection is essential to prevent pathogen carryover.

Incidence of Horse Manure Compost on Human and Horse Health Human Health

Any human in contact with an infected animal or disease vector is potentially at risk of contracting a zoonotic disease. However, whereas immunocompetent persons generally have a smaller risk of contracting a horse manure related infection, the following persons remain at risk:

- babies and infants (immature immune systems, poor hygiene - hands in mouth, etc.)
- asthma and allergy sufferers
- pregnant women (immune systems are more susceptible and there are additional fetal hazards)
- the elderly (immune systems may be impaired)
- immunocompromised people (undergoing cancer therapy, HIV/AIDS patients, etc.)
- individuals taking oral or intravenous antibiotics.

Due to their higher frequency of exposure to horse manure, animal healthcare workers and veterinarians could also be considered a population at risk.

While the list of populations at risk seems quite long, to our knowledge the literature has only reported two human infections related to horse manure: (*i*) a severe *Streptococcus zooepidemicus* infection contracted by a 79-year-old gardener who denied exposure to animals but was an avid gardener using horse manure as a fertilizer [21] (*ii*) a primary subcutaneous abscess caused by *Rhodococcus equi* contracted by a 65-year-old heart transplant recipient who collected fresh horse manure for his garden daily from a local paddock and reported no direct contact with animals or ingestion of unpasteurized dairy products [22]. In both cases, infections were related to non-composted horse manure. To our knowledge, there is no reported human infection related to horse manure compost. However, even if the number of horse manure related infections is low, the establishment of the epidemiologic origin of an infection is sometimes difficult and the number of infections associated with horse manure or compost could be underestimated. It would be very difficult to determine the origin of an infection in the case, for example, of manure spread over land leading to the introduction of zoonotic pathogens into soil that may subsequently contaminate vegetables [23] or water [2]. From 1993 to 1994, one third of the outbreaks associated with drinking water were due to *Giardia spp.* and *C. parvum* [24], for example, and we could not exclude the possibility of some oocysts originally coming from horse manure.

Horses

Logically, the life cycle of pathogens found in horse manure is orofecal. The fact that horses are infected by these pathogens implies the existence of a link between manure and horse food and/or water supplies. The fact that many horses are affected by diarrhea and parasitic diseases shows that this problem is therefore of great concern and must be taken into account in horse breeding. In this context, it seems difficult to assess the problem of the influence of pathogen inactivation during horse manure composting on horse health given that it is almost impossible to determine if an infection is related to horse manure or to horse manure compost. It has already been shown that farms fertilizing pastures with horse manure had a significantly higher risk of being *Parascaris equorum* fecal egg count positive [8, 25] but, to our knowledge, there is no study reporting a direct link between a horse infection and the spreading of horse manure compost. Additional studies are necessary to determine the influence of horse manure composting on horse health.

CONCLUSION

In 1908, Harold Bolce argued in Appleton's Magazine in an article entitled "The Horse *vs* Health" that most of the modern city's sanitary and economic problems were caused by the horse. Bolce charged that each year 20,000 New Yorkers died from "maladies that fly in the dust, created mainly by horse manure". Nowadays, the risk of contracting a horse-related infection is far smaller, so it is quite rare to find in the literature cases of human infections related to horse manure and/or compost. On the other hand, as numerous equine diseases are related to an orofecal cycle, it is essential that the use of composted horse manure not contribute to the transmission of these pathogens.

When the composting process is carried out correctly (*e.g.* when it complies with US-EPA recommendations), it is generally established that pathogens should be properly inactivated. Theoretically, then, the utilization of composted horse manure should not represent a concern for horse or human health. However, due to the many factors that could influence the time–temperature course for a particular mixture of materials under specific environmental conditions (many of which cannot be measured, let alone controlled), it appears very difficult to predict the effectiveness of pathogen inactivation, and thus pathogen population evolution, during a given composting process. Moreover, parameters like compost heterogeneity and pathogen regrowth make it practically impossible to ensure the biological safety of an entire horse manure compost pile. Finally, even if the analysis of the bacterial and parasitic content of a compost sample does not reveal pathogen contamination, it is not practicable to ensure that the entire compost pile is pathogen-free. Consequently, it should be recommended to place compost in an area away from other animals where runoff will not contaminate the water supply, and to avoid spreading horse manure compost in fields where horses graze even if time-temperature requirements have been satisfied.

In addition to the initial problem of manure composting itself, it appears just as important to have good hygiene practices at the same time, with daily cleaning and regular disinfection of horse boxes, for example [8]. Particular care should also be taken with young foals that, on the one hand are particularly susceptible to infections, and on the other, constitute an important source of pathogen shedding.

REFERENCES

[1] Lundgren, J; Pettersson, E. Combustion of horse manure for heat production. *Bioresource Technol*, 2009, 100(12), 3121-6.

[2] Pell AN. Manure and microbes: public and animal health problem? *J Dairy Sci*, 1997, *October*, 1, 80(10), 2673-81.

[3] Krogmann, U; Westondorf, ML; Rogers, BF. Best management practices for horse manure composting on small farms. *Rutgers Bulletin.*

[4] EPA, US. Control of pathogens and vector attraction in sewage sludge. 2003, EPA/625/R-92/013, 173-7.

[5] Wichuk, KM; McCartney, D. A review of the effectiveness of current time-temperature regulations on pathogen inactivation during composting. *J Environ Eng Sci*, 2007, 6(5), 573-86.

[6] Krogstad, O; Gudding, R. The survival of some pathogenic microorganisms during reactor composting. *Acta Agric Scand*, 1975, 25, 281–4.

[7] Elschner, M; Schrader, C; Hotzel, H; Prudlo, J; Sachse, K; Eichhorn, W; et al. Isolation and molecular characterisation of equine rotaviruses from Germany. *Vet Microbiol*, 2005, 105(2), 123-9.

[8] Fritzen, B; Rohn, K; Schnieder, T; von Samson-Himmelstjerna, G. Endoparasite control management on horse farms - lessons from worm prevalence and questionnaire data. *Equine Vet J*, 2010 Jan; 42(1), 79-83.

[9] Johnson, E; Atwill, E; Filkins, M; Kalush, J. The prevalence of shedding of *Cryptosporidium* and *Giardia spp.* based on a single fecal sample collection from each of 91 horses used for backcountry recreation. *J Vet Diagn Invest*, 1997, January 1, 1997, 9(1), 56-60.

[10] Olson, ME; Thorlakson, CL; Deselliers, L; Morck, DW; McAllister, TA. *Giardia* and *Cryptosporidium* in Canadian farm animals. *Vet Parasitol*, 1997, 68(4), 375-81.

[11] Current, WL. The biology of *Cryptosporidium*. A. S. M. *News*, 54, 605. 1988.

[12] Russell, AD. Lethal effects of heat on bacterial physiology and structure. *Sci prog*, 2003, 86(1-2), 115-37.

[13] Guardabassi, L; Dalsgaard, A; Sobsey, M. Occurrence and survival of viruses in composted human faeces. Sustainable Urban Renewal and Wastewater Treatment No. 32, Danish Environmental Protection *Agency, Danish Ministry of the Environment.*, 2003.

[14] Nell, JH; Steer, AG; van Rensburg, PAJ. Hygienic quality of sewage sludge compost. *Water Sci Technol*, 1983, 19(181-194).

[15] Droffner, ML; Brinton, WF. Survival of *E. coli* and *Salmonella* populations in aerobic thermophilic composts as measured with DNA gene probes. *Zentralbl Hyg* Umweltmed, 1995, Jun;197(5), 387-97.

[16] Steer, AG; Windt, CN. Composting and fate of *Ascaris lumbricoides* ova. *Water SA*, 1978, 4(129-133).

[17] Burge, WD; Cramer, WN; Epstein, E. Destruction of pathogens in sewage sludge by composting. *Trans ASAE*, 1978, 21(510-544).

[18] Pereira-Neto, JT; Stentiford, EI; Mara, DD. Low cost controlled composting of refuse and sewage sludge. *Water Sci Technol*, 1987, 19(839–845).

[19] Strader, R; Bromhal, GS. Evaluation of temperature fluctuations in a windrow composting operation. Proceedings of the 1997 Air & Waste Management Association's 90th Annual Meeting & Exhibition, Jun 8-13 1997, Toronto, Ont 10 1997.

[20] Soares, HM. P*athogen indicator regrowth potential as a method to evaluate compost stability*. Ph.D. Thesis. University of Massachusetts Amherst., 1996.

[21] Lee, AS; Dyer, JR. Severe *Streptococcus zooepidemicus* infection in a gardener. *Med J Aust*, 2004, Apr 5, 180(7), 366.

[22] Adal, KA; Shiner, PT; Francis, JB. Primary subcutaneous abscess caused by *Rhodococcus equi*. *Ann Intern Med*, 1995, February 15, 1995, 122(4), 317.

[23] Islam, M; Morgan, J; Doyle, MP; Phatak, SC; Millner, P; Jiang, X. Fate of *Salmonella enterica s*erovar *Typhimurium* on carrots and radishes grown in fields treated with contaminated manure composts or irrigation water. *Appl Environ Microbiol*, 2004, April 1, 2004, 70(4), 2497-502.

[24] Kramer, MH; Herwaldt, BL; Craun, GF; Calderon, RL; Juranek, DD. Waterborne disease: 1993 and 1994. *J Am Water Works Assoc,* 1996, 88(3), 66-80.

[25] Lyons, ET; Tolliver, SC; Drudge, JH. Historical perspective of cyathostomes: prevalence, treatment and control programs. *Vet Parasitol*, 1999, Aug 31, 85(2-3), 97-111.

[26] Hathcock, TL; Schumacher, J; Wright, JC; Stringfellow, J. The prevalence of *Aeromonas* species in feces of horses with diarrhea. *J Vet Intern Med*, 1999, Jul-Aug, 13(4), 357-60.

[27] Netherwood, T; Wood, JL; Townsend, HG; Mumford, JA; Chanter, N. Foal diarrhoea between 1991 and 1994 in the United Kingdom associated with *Clostridium perfringens*, rotavirus, *Strongyloides westeri* and *Cryptosporidium spp. Epidemiol Infect*, 1996, Oct, 117(2), 375-83.

[28] Frederick, J; Giguère, S; Sanchez, LC. Infectious agents detected in the feces of diarrheic foals: a retrospective study of 233 cases (2003-2008). *J Vet Intern Med*, 2009, 23(6), 1254-60.

[29] Wilkins, CA; Richter, MB; Hobbs, WB; Whitcomb, M; Bergh, N; Carstens, J. Occurrence of *Clostridium tetani* in soil and horses. *S Afr Med J*, 1988, Jun 18, 73(12), 718-20.

[30] Johnson, AL; McAdams, SC; Whitlock, RH. Type A botulism in horses in the United States: a review of the past ten years (1998-2008). *J Vet Diagn Invest March 1*, 2010, 22(2), 165-73.

[31] Golomidova, A; Kulikov, E; Isaeva, A; Manykin, A; Letarov, A. The diversity of coliphages and coliforms in horse feces reveals a complex pattern of ecological interactions. *Appl Environ Microbiol*, 2007 October 1, 2007, 73(19), 5975-81.

[32] Gudmundsdottir, KB; Svansson, V; Aalbak, B; Gunnarsson, E; Sigurdarson, S. *Listeria monocytogenes* in horses in Iceland. *Vet Rec*, 2004 October 9, 2004, 155(15), 456-9.

[33] Muscatello, G; Leadon, DP; Klayt, M; Ocampo-Sosa, A; Lewis, DA; Fogarty, U; et al. *Rhodococcus equi* infection in foals: the science of 'rattles'. *Eq Vet J*, 2007, Sep;39(5), 470-8.

[34] Weaver, RW; Entry, JA; Graves, A. Numbers of fecal streptococci and *Escherichia coli* in fresh and dry cattle, horse, and sheep manure. *Can J Microbiol*, 2005, Oct;51(10), 847-51.

[35] Corning, S. Equine cyathostomins: a review of biology, clinical significance and therapy. *Parasites & Vectors*, 2009, 2, (Suppl 2), S1.

[36] Getachew, M; Trawford, A; Feseha, G; Reid, S. Gastrointestinal parasites of working donkeys of Ethiopia. *Trop Anim Health Pro*, 2010, 42(1), 27-33.

[37] Chapman, MR; French, DD; Klei, TR. Seasonal transmission of gastrointestinal parasites of equids in southern Louisiana. *J Parasitol*, 2001, Dec;87(6), 1371-8.

[38] Hébert, L; Cauchard, J; Doligez, P; Quitard, L; Laugier, C; Petry, S. Viability of *Rhodococcus equi* and *Parascaris equorum* eggs exposed to high temperatures. *Current Microbiology*, 2010, 60, 38-41.

[39] Veronesi, F; Diaferia, M; Piergili Fioretti, D. *Anoplocephala perfoliata* infestation and colics in horses. *Vet Res Commun*, 2009, 33(0), 161-3.

[40] Brinton, WF. Compost quality standards & guidelines *Woods End Research Laboratory*, Inc 2000.

In: Horses: Biology, Domestication, and Human Interactions ISBN: 978-1-61761-303-6
Editor: James E. Leffhalm

Chapter 6

EQUINE NATIVE MICROBIOTA AS A SOURCE OF BENEFICIAL MICROBES

Martín Fraga[1*], Karen Perelmuter[1], Luis Delucchi[2b] and Pablo Zunuino[1]

[1]Departamento de Microbiología, Instituto de Investigaciones Biológicas Clemente Estable, Avenida Italia 3318, CP 11600 Montevideo, Uruguay
[2] Facultad de Veterinaria, Universidad de la República, Uruguay, Lasplaces 1550, Montevideo, Uruguay

ABSTRACT

Indigenous microbiota is a complex community of microorganisms that colonizes the whole gastrointestinal tract, some parts of other tracts and the air-exposed surfaces of animals. Generally, this community confers protection against the onset of pathogen infection and also provides many others benefits such as immune modulation or production of vitamins. Gastrointestinal microbiota is by far the best characterized microbiota in different animal species and in the human being but there is an increasing interest in other groups of microorganisms, such as those associated to the genitourinary tract.

Composition of these microbial communities varies between species, individuals and even within the different parts of a single organ. For example, the microorganisms of different portions of the enteric tract are strikingly different. Now it is also known that there is a host and site specific microbiota composition, also related to other factors such as age or health condition.

Probiotics are live microorganisms that when administered in adequate amounts confer a health benefit to the host. Preferably they should be native isolates, so the research in native microbiota acquires a significant relevance since they can be sources of potential beneficial microorganisms.

So far, equine native microbiota has not been extensively characterized. These microbial communities are related to the normal functions of different organisms and its

* Corresponding author: E-mail: pablo@iibce.edu.uy

characterization may provide important data about beneficial microbes useful for the promotion of horses' health.

In equines, probiotics could be used with different purposes, including improvement of the diet digestibility and prevention of metabolic or infectious disorders. In mares, probiotics could improve milk quality and quantity and microorganisms isolated from the vaginal tract could be directed to the prevention of fertility problems related to endometriris, a serious reproductive problem in the horse. Probiotics for foals would be useful for growth promotion and control of the problems associated with diarrheas. Also, probiotics-based strategies could be designed for specific groups of animals, such as racing horses.

In this chapter, equine gastrointestinal and vaginal microbiota are reviewed and attention is focused on modulation strategies based on the use of probiotics for health promotion.

1. Equine Native Microbiota

Gastrointestinal microbiota has been by far the most studied microbiota in several animal species and recently, attention has been set also into vaginal microbiota, particularly in the human being.

The natural pattern of feeding and digestion of horses has been disturbed by domestication with concomitant frequency of gastrointestinal diseases, which became the major cause of morbidity and the most important cause of death (Daily et al. 2001). Normal and healthy microbiota is important for the wellbeing of the intestine of most animal species. Therefore, the knowledge of the equine microbial community will shed light in the prevention, treatment and understanding the development and setting of diseases.

Genital infections can lead to fertility problems and abortion (Bjurström 1993). Particularly, endometritis is a cause of fertility loss in mares, provoking important losses to the equine industry (Brook 1984). Many plausible uterine pathogens are habitants of the vagina. The knowledge of the bacterial biota inhabiting the genital tract of the mare would be important to develop therapeutic strategies to control the setting of disease.

1.1. Gastrointestinal Tract

The gastrointestinal tract (GIT) of horses, as in other species, can be divided in three different portions: foregut, midgut and hindgut (Stevens et al. 1998). The first portion of the tract comprises the esophagus and stomach, midgut is formed by duodenum and jejunum-ileum and the hindgut, by far the most studied region of the equine GIT, comprises the cecum, colon and rectum.

The stomach of horses bears a non-secreting area, the non-glandular stomach, and a secreting area, the glandular stomach. This organ is small if we consider other large animal species, with a capacity of 7.5 to 15 L (Al Jassim et al. 2009). The stomach microbial community is relevant because of its role in the inoculation of food particles and the consequent development of microbial communities in the lower regions of the GIT. Plausible GIT probiotics and GIT pathogens should be able to survive and inoculate food in this region. However, microbiota associated to this organ so far has almost not been studied.

de Fombelle and colleagues observed that total anaerobic bacteria from stomach contents are similar in slaughtered animals fed with pellets rich in fiber and straw in French stables habits, and in typical performance horses fed with pelleted food rich in cereal and meadow hay (de Fombelle et al. 2003). Anaerobic bacterial counts in those conditions ranged from 7.45×10^7 to 3.75×10^9 colony-forming units per mL (CFU/mL). Similar numbers could be found in conscious animals which stomach contents samples were obtained by naso-gastric intubation (4.78×10^6 to 7.07×10^8 CFU/mL, Varloud et al. 2007).

Lactobacilli populations counts were influenced by different diets. A least square mean of 6.30×10^7 CFU/mL was reported for French habits diet and a lest square mean of 2.5×10^8 CFU/mL for horses fed with pelleted food rich in cereal and meadow hay (de Fombelle et al. 2003). In addition, Varloud et al. (2007) found a significant variation between individuals (n=6) with a range of 1.41×10^6 to 7.76×10^6 CFU/mL.

Yuki et al. (2000) studied the colonization by indigenous lactobacilli of the stratified squamous epithelium of the non-secreting area of a euthanized mare. Authors described a dense layer of Gram positive rods on the keratinized stratified squamous epithelium and reported more than one hundred bacteria attached per epithelial cell. The *Lactobacillus* species present in the stomach of the horse are *Lactobacillus reuteri, Lactobacillus salivarus, Lactobacillus crispatus, Lactobacillus agilis, Lactobacillus mucosae,* and *Lactobacullis delbrueckii* (Al Jassim et al. 2005; Yuki et al. 2000). *L. salivarus* can be found in different parts of the GIT and it is reported to be adherent to the epithelial cells of the stomach (Yuki et al. 2000). Another relevant finding was that the highest concentration of lactate-utilizing bacteria was detected in the stomach (de Fombelle et al. 2003).

In the last few years, culture-dependant and molecular approaches have shown a more diverse stomach bacterial community including species belonging to genera such as *Lactobacillus*, *Streptococcus*, *Clostridium*, *Prevotella*, *Pseudomonas*, and *Propionibacterium* as the predominant microbial groups and *Escherichia*, *Legionella*, *Moraxella*, *Rhodococcus*, *Veillonella*, *Eubacterium* and *Pasteurella* as less prevalent members (Al Jassim et al. 2008).

The small intestine of the horse comprises the duodenum, jejunum and ileum. This is the major site of chemical digestion and absorption of aminoacids, simple sugars and fatty acids (Al Jassim et al. 2009). Grass-fed horses harbor an anaerobic bacterial population of about 2.9×10^6 UFC/g in duodenum, 2.9×10^7 UFC/g in jejunum and 3.8×10^7 UFC/g in ileum. In animals under typical French stables habits and in performance horses counts were 4×10^8 and 5×10^8 UFC/mL respectively in jejunum and 4×10^7 and 5×10^7 UFC/mL in ileum under the different diets and housing conditions (Mackie et al. 1988; de Fombelle et al. 2003). In spite of the diet variation, lactobacilli counts are higher in ileum than in other GIT areas. Performance horses harbor a more abundant lactobacilli population than those with French stables habits (de Fombelle et al. 2003).

The cecum, colon and the rectum form the large intestine of the horse in which a symbiotic community of anaerobic microorganisms including bacteria, fungi, protozoa and archae lives (Daly et al. 2001). This complex community hydrolyzes plant fiber and produces short chain fatty acids providing 60-70% of horse body energy. Bacteria community inhabiting the horse hindgut resembles in function and composition the one of ruminant animals (Stevens et al. 1998). Attempts to study this community has been more frequent in literature than with others parts of the tract and authors had dealt with samples such as feces and intestinal contents to describe the hindgut populations. However, fecal samples seem to provide insufficient means to monitor changes in cecal bacteria during the laminitis process. (Milinovich et al. 2008).

Bacterial community in the cecum and the colon of grass-fed animals can account numbers of 2.05 x10^9 CFU/g and 1.26 x10^9 CFU/g respectively (Mackie et al. 1988), and 3.63 x10^8 CFU/g in feces of animals not suffering intestinal disorders, (Garrett et al. 2002). Lactobacilli counts in animals fed only with a high energy forage were 7.08 x10^5-7.08 x10^6 bacteria per gram of feces while in animals having barley hay or high energy forage both supplemented with a grain based concentrate numbers were at least one order higher (Willing et al. 2009; Endo et al. 2009).

Some studies have focused in identifying and detecting lactic acid bacteria (LAB) and related microorganisms such as *Bifidobacterium* spp. because of their potential use as probiotics and their possible relation with pathologies such as laminitis. *Lactobacillus equi, L. salivarus, L. mucosae* and *L. agilis* can be detected under different feeding habits (Morotomi et al. 2002; Al Jassim et al. 2005; Endo et al. 2009). *Lactobacillus ruminis* was only detected in animals fed with diets supplemented with grain concentrate (Willing et al. 2009) and also *Lactobacillus johnsonii, Lactobacillus hayakitensis, Lactobacillus equigenerosi, Weissela confusa/cibaria, Parascardovia denticolen, Bifidobacterium pseudolongum* (Al Jassim et al. 2005; Morita et al. 2007; Wiiling et al. 2009); *Mitsuokella jalaudinii* and *Veillonella atypica* were detected in horses with induced laminitis (Al Jassim et al. 2005). The *Streptococcus bovis-Streptococcus equinus* complex could be detected and isolated in many different situations and from different parts of the GIT. Their relative abundance is increased and reaches values of 50% and more in feces of horses experimentally induced laminitis (Al Jassim et al. 2005; Milinovich et al. 2007). *Lactobacillus pentosus* was isolated from faces of a clinically healthy horse and its probiotic potential in vitro was assayed (Weese et al. 2004).

In animals suffering from equine grass sickness (EGS) bacterial counts in feces for anaerobes and aerobes can be 10 to 100-fold higher than those found in healthy animals. Among the different types of Gram positive rods bacteria are significantly higher in chronic cases of EGS than in control (non-affected) animals. Many clostridial species were described in animals with EGS, such as *Clostridium barati, Clostridium bifermentans, Clostridium botulinum* type C/D, *Clostridium carnis, Clostridium clostridioforme, Clostridium difficcile, Clostridium glycolicum, Clostridium limosum, Clostridium perfringens, Clostridium sardiniensis, Clostridium sordelli, Clostridium sphenoides, Clostridium sporogenes* and *Clostridium tertium. C. bifermentans* was also found in a healthy animal and was the only clostridial described (Garrett et al. 2002). *Peptostreptococcus anaerobius, Peptostreptococcus productus, Peptostreptococcus prevotii, Peptostreptococcus micros* and *Sarcinia ventriculi* were found both in healthy and in horses suffering from gastrointestinal disease (Garrett et al. 2002).

Daly and colleagues reported one of the first molecular approaches attempting to study the microbial diversity within the equine GIT (Daly et al. 2001). These authors constructed 16S rRNA gene libraries from intestinal contents of grass-fed animals freshly slaughtered and genetic diversity in equine cecum and colon was described. Low GC Gram positive bacteria (LGCGP) were 72% of the total recovery of a clone population. This would imply a high representation of these bacteria in this community. In this phylum are included some cellulolytic and fibrolytic bacteria such as *Clostridiuim* spp., *Ruminococcus* spp., *Butyrivibrio* spp. and *Eubacterium* spp. Only the 11% of the recovered sequences were more than 97% similar with published sequences in databases while Willing et al. (2009) could detect about

27% of known sequences. Among these sequences only 5% could be affiliated to previously known species such as *S. bovis, Ruminococcus flavefaciens, Clostridium barati, Butyrivibrio fibrosolvens, L. salivarus* and *Campylobacter lanienae* (Daly et al. 2001). Interestingly, a recent study performed with feces of Hokkaido native and light horses that had been grazed on snow-covered bamboo grass showed different patterns of the major phylum represented. In these animals, LGCGP represented between 36.8 and 55.3% while the major phylum was Cytophaga-Flavobacter-Bacteroides accounting between 44.7 and 47.4% (Yamano et al. 2008) in contrast with the 20% of this phylum reported by Daly and coworkers (2001).

1.2. Vaginal Microbiota

Vaginal microbiota of vertebrates is a complex consortia of microorganisms in which LAB play a fundamental role. In humans, the presence of LAB and particularly lactobacilli is considered a health-disease indicator and its predominant place in this ecological niche is well recognized (Redondo-Lopez et al. 1990). However, little is known about mare vagina microecology. Mare vaginal microbiota has been studied attempting to characterize pathogens. Pathogens organisms isolated from the vagina were *Corynebacterium* sp., *Streptococcus* sp. and *Staphylococcus* sp. (Newcombe 1978; Hinrichs et al. 1988).

In a study in which native equine vaginal LAB were analyzed, 26 healthy mares were sampled and the abundance of LAB and *Lactobacillus* spp. were evaluated. Also, a group of isolates was obtained and characterized in order to assay their potential as probiotics (Fraga et al. 2008). Mares were distributed in two populations. One of them consisted in mixed-bred family working horses with poor stabling and variable feeding conditions while the other included only pure-bred horses maintained in good housing and controlled feeding conditions. LAB were detected in all vaginal samples but bacterial counts were very heterogeneous among the different individuals ranging from 31,6 to 1 x 10^8 CFU per vaginal swabbing. Lactobacilli growth was observed in 70% of the samples with a median of 2.5 $\log_{10}$ CFU and a maximum of 5.9 $\log_{10}$ CFU per vaginal swabbing. Interestingly, a significant correlation between lactobacilli counts and horse habits was found. Significant higher vaginal lactobacilli counts were found in mixed-bred family working horses under poor stabling conditions compared to those detected in animals under good housing and controlled feeding. Authors suggest that these horses may have restricted chances to acquire a varied indigenous lactic biota.

Isolates from this study were identified as *Lactobacillus pantheris, L. equi, L. mucosae, Enterococcus faecalis* and *Enterococcus faecium.* The detection of these organisms in the mare vagina suggests plausible intestine vagina transference of the microbiota due to the presence of these organisms in the intestinal ecosystem (Al Jassim et al. 2005; Morotomi et al. 2002; Franz et al. 1999; Morelli et al. 2004).

2. PROBIOTICS

Probiotics were first considered as microorganisms that could contribute to the balance of the intestinal microbiota through oral administration (Fuller 1989). Today this definition has widen and now probiotics are considered as live microorganisms which when administered in adequate amounts confer a health benefit on the host (FAO/WHO 2001).

The possibility to use this kind of feed supplements to improve animal health, welfare and productivity through manipulation of the native microbial ecosystems has gained considerable attention in the last decades. Growth-promoting antimicrobials have been widely distributed and are still used in several countries.

However, there are increasing safety concerns about the risk of releasing antibiotic resistance in the environment and favoring the rise of bacterial resistance, and the persistence of chemical residues in animal products (Chaucheyras-Durand et al. 2010). For example, since January of 2006, the European Union banned the feeding of all antibiotics and related drugs to livestock for growth promotion purposes.

Therefore, alternative strategies based on supplementation of 'natural' products such as probiotics, have been developed to improve animal health and productivity. Increasing amounts of scientific data are supporting the view that these products, which are defined as a source of live (viable) naturally occurring microorganisms, can beneficially affect the balance of GIT microbiota and that they have a real benefit in animal nutrition and health.

The use of native probiotic isolates is preferably than exogenous ones, since native strains would be better adapted to the target ecological niche (O´Sullivan 2001). As an example, larger inoculums of exogenous bacteria are needed to achieve similar results than with native (Vinderola et al. 2004). Therefore, the research in indigenous microbiota acquires a significant relevance since these communities can become sources of potential beneficial microorganisms.

3. Probiotics for Horses

In equines, probiotics could be used with different purposes, including improvement of the diet digestibility and prevention of metabolic or infectious disorders. In mares, probiotics could improve milk quality and quantity and microorganisms isolated from the vaginal tract could be directed to the prevention of fertility problems related to endometritis, a serious reproductive problem in the horse. Probiotics for foals would be useful for growth promotion and control of the problems associated with diarrheas. In addition, probiotics-based strategies could be designed for the health promotion of specific groups of animals, such as racing horses, which are exposed to stress caused by long transportation periods and racing or high intake of rich concentrate diet (Chaucheyras-Durand et al. 2010).

Probiotics have been used in horses for therapy of digestive tract disease, for example for the treatment of *Clostridium difficile* associated diarrhea. However, the use in foals of native *Lactobacillus pentosus* WE7 to prevent diarrheas showed poor results (Weese et al. 2005). These results could be due to the improper selection of the organism, so further studies using alternative strains should be performed.

On the other hand, a randomized double blinded clinical trial conducted to estimate the incidence of *Salmonella* fecal shedding in hospitalized horses administered a probiotic versus a placebo showed promising results. The authors reported that administration of a commercial probiotic reduced the incidence of *Salmonella* shedding after 48 hours of hospitalization by about 65 % (Ward et al. 2004).

In horses, the energy supplied by forage diets frequently does not meet the energy requirements for intense physical exercise due to the relatively low digestibility of cell wall carbohydrates. Consequently, in animals subjected to strong efforts conditions, diets are supplemented with concentrate-based diets resulting in ingestion of large quantities of starch by animals over a short time (Jouany et al. 2008). The accumulation of lactate and the subsequent pH decrease, affects the activity of the cellulolytic bacterial populations and increases the susceptibility of horses to colic pain or laminitis (Kronfeld et al. 1997).

Laminitis, considered the most severe disease of the equine foot, occurs when the attachment between the distal phalanx and inner hoof wall fails (Pollitt 1996). Failure of the attachment apparatus of the foot results in the distal phalanx being driven downwards under the horse's weight, damaging the vasculature of the hoof and crushing the corium of the sole and coronet, resulting in lameness (Pollitt 1996). The pathology of acute laminitis and its association with lactic acidosis following carbohydrate overload has been well described (Garner et al. 1977) but its pathogenesis is not well elucidated. The exposure to excess of carbohydrates results in change in the hindgut microbiota. The pathology of the hoof disease is associated with this change on the microbiota probably by lamellar ischemia for the hindgut endotoxins or activation of matrix metalloproteinases or lactic acidosis itself. Oligofructose-induced laminitis showed a *S. bovis-S. equinus* complex proliferation in feces before the onset of laminitis (Milinovich et al. 2007).

Swyers and colleagues did not obtain satisfactory results when they used a commercial mixture of LAB to increase digestibility and minimize the risk of acidosis in horses (Swyers et al. 2008). The design of probiotics-based strategies, directed to the use of lactate-consuming bacteria rather than LAB could be useful for the prevention of acidosis related to a high starch intake.

CONCLUSION

The study of indigenous microbiota in humans and other animal species has gained an increasing importance, since these communities are crucial for the normal functions of the organisms. In addition, this microbiota has a huge potential for the finding of microbes or even genes with medical, biotechnological or even industrial importance. Specifically in horses, there are few studies on these issues. Therefore, more research will be needed to study these communities including the finding of strains for their use as probiotics.

REFERENCES

Al Jassim, R; Andrews, FM. The bacterial community of the horse gastrointestinal tract and its relation to fermentative acidosis, laminitis, colic and stomach ulcers. *Vet Clin Equine*, 2009, 25, 199-215.

Al Jassim, R; Mc Gowan, T; Andrews, F; Mc Gowan, C. *Gastric ulceration in horses, the role of bacteria and lactic acid.* Rural Industries Research and Develpment Corporation, 2008 Publication No. 08/033. www.rirdc.gov.au.

Al Jassim, R; Scott, PT; Trebbin, AL; Trott, D; Pollitt, CC. The genetic diversity of lactic acid producing bacteria in the equine gastrointestinal tract. *FEMS Microbiol Lett*, 2005, 248, 75-82.

Bjurström, L. Aerobic bacteria occurring in the vagina of bitches with reproductive disorders. *Acta Vet Scand*, 1993, 34, 29-34.

Brook, D. Diagnosis of equine bacteria endometritis. *Compend Contin Educ Vet Pract*, 1984, 6, 300-307.

Chaucheyras-Durand, F; Durand, H. Probiotics in animal nutrition and health. *Beneficial Microbes*, 2010, 1, 3-9.

Daly, K; Stewart, CS; Flint, HJ; Shirazi-Beechey, SP. Bacterial diversity within the equine large intestine as revelead by molecular analysis of clone 16S rRNA genes. *FEMS Microbiol Ecol*, 2001, 38, 141-151.

de Fombelle, A; Varloud, M; Goachet, A; Jacotot, E; Philippoeau, C; Drogoul, C; Julliand, V. Characterization of the microbial and biochemical profile of the different segments of the digestive tract in horses given two distinct diets. *Anim Sci*, 2003, 77, 293-304.

Endo, A; Futagawa-Endo, Y; Dicks, LMT. *Lactobacillus* and *Bifidobacterium* diversity in horse feces revealed by PCR-DGGE. *Curr Microbiol*, 2009, 59, 651-655.

FAO/WHO. Joint FAO/WHO expert consultation on evaluation of health and nutritional properties of probiotics in food. Córdoba, Argentina. 2001.

Fraga, M; Perelmuter, K; Delucchi, L; Cidade, E; Zunino, P. Vaginal lactic acid bacteria in the mare: evaluation of the probiotic potential of native *Lactobacillus* spp. and *Enterococcus* spp. strains. *Antonie van Leeuwenhoek*, 2008, 93, 71-78.

Franz, C; Holzapfel, WH; Stiles, M. Enterococci at the crossroads of food safety. *Int J Food Microbiol*, 1999, 47, 1-24.

Fuller, R. Probiotics in man and animals. *J Appl Bacteriol*, 1989, 66, 365-378.

Garner, HE; Hutcheson, DP; Coffman, JR; Hahn, AW; Salem, C. Lactic acidosis: a factor associated with equine laminitis. *J Anim Sci*, 1977, 45, 1037-1041.

Garret, LA; Brown, R; Poxton, IR. A comparative study if the intestinal microbiota of healthy horses and those suffering from equine grass sickness. *Vet Microbiol*, 2002, 87, 81-88.

Hinrichs, K; Cumings, MR; Sertich, PL; Kenney, RM. Clinical significance of aerobic bacterial flora of the uterus, vagina, vestibule and clitoral fossa of clinically normal mares. *Javma*, 1988, 193, 72-75.

Jouany, JP; Gobert, J; Medina, B; Bertin, G; Julliand, V. Effect of live yeast culture supplementation on apparent digestibility and rate of passage in horses fed a high-fiber or high-starch diet. *J Anim Sci*, 2008, 86, 339-347.

Kronfeld, DS; Harris, P. Feeding the athletic horse. In: KN; Thompson, editor. The veterinarian's practical reference to equine nutrition. St Louis, MO: Purina Mills Inc & AAEP; 61-77.

Mackie, R; Wilkins, C. Enumeration of anaerobic bacteria microflora of the equine gastrointestinal tract. *Appl Environ Microbiol*, 1988, 54, 2155-2160.

Millinovich, GJ; Trott, DJ; Burrell, PC; Croser, EL; Al Jassim, RAM; Morton, JM; van Eps, AW; Pollitt, CC. Fluorescent in situ hybridization analysis of hindgut bacteria associated with the development of equine laminitis. *Environ Microbiol*, 2007, 9, 2090-2100.

Morelli, L; Zonenenschain, D; Del Piano, M; Cognein, P. Utilization of the intestinal tract as a delivery system for urogenital probiotic. *J Clin Gastroenterol*, 2004, 38, 107-110.

Morita, H; Shiratori, C; Murakami, M; Takami, H; Kato, Y; Endo, A; Nakajima F; Takagi M; Akita, H; Okada, S; Masaoka, T. *Lactobacillus hayakitensis* sp. nov., isolated from intestines of healthy thoroughbreds. *Int J Syst Evol Microbiol*, 2007, 57, 2836-2839.

Morotomi, M; Yuki, N; Kado, Y; Kushiro, A; Shimazaki, T; Watanabe, K; Yuyama, T. *Lactobacillus equi* sp. nov., a predominant intestinal *Lactobacillus* species of the horse isolated from faeces of healthy horses. *Int J Sys Evol Microbiol*, 2002, 52, 211-214.

Newcombe, JR. Comparison of the bacterial flora of three sites in the genital tract of the mare. *Vet Rec*, 1978, 102, 169-70.

O'Sullivan, GC. Probiotics. *Brit J Surg,* 2001, 88, 161-162.

Pollitt, CC. Basement membrane pathology: a feature of acute equine laminitis. *Equine Vet J*, 1996, 28, 38-46.

Redondo-López, V; Cook, RL; Sobel, JD. Emerging role of lactobacilli in the control and maintenance of the vaginal bacterial microflora. *Rev Infect Dis*, 1990, 12, 856-872.

Stevens, CE; Hume, ID. Contributions of microbes in vertebrate gastrointestinal tract to production and conservation of nutrients. *Physol Rev*, 1998, 78, 393-427.

Swyers, KL; Burk, AO; Hartsock, TG; Ungerfeld, EM; Shelton, JL. Effects of direct-fed microbial supplementation on digestibility and fermentation end-products in horses fed low- and high-starch concentrates. *J Anim Sci*, 2008, 86, 2596-2608.

Varloud, M; Rousell, A., Boisot, P; Julliand, V. A technique for the collection and the study of biochemical and microbial characteristics of postprandial gastric contents from conscious horses. *Anim Feed Sci Technol*, 2007, 133, 259-274.

Vinderola, CG; Mocchiutti, P; Reinheimer, JA. Interactions among lactoc acid starter and probiotic bacteria used for fermented dairy products. *J Dairy Sci*, 2002, 85, 721-729.

Ward, MP; Alinovi, CA; Couëtil, LL; Glickman, LT; Wu, CC. A randomized clinical trial using probiotics to prevent Salmonella fecal shedding in hospitalized horses. *J Eq Vet Sci*, 2004, 24, 242-247.

Weese, JS; Anderson, MEC; Lowe, A; Penno, R; Da Costa, TM; Button, L; Goth, KC. Screening of the equine intestinal microflora for potential probiotic organisms. *Equine Vet J*, 2004, 36, 351-355.

Weese, JS; Rousseau, J. Evaluation of *Lactobacillus pentosus* WE7 for prevention of diarrhea in neonatal foals. *J Am Vet Med Assoc*, 2005, 226, 2031-2034.

Willing, B; Vörös, A; Roos, S; Jones, C; Jansson, A; Lindberg, JE. Changes in faecal bacteria associated with concentrate and forage-only diets fed to horses in training. *Equine Vet J*, 2009, 41, 908-914.

Yamano, H; Koike, S; Kobayashi, Y; Hata, H. Phylogenetic analysis of hindgut microbiota in Hokkaido native horses compared to light horses. *Anim Sci*, 2008, 79, 234-242.

Yuki, N; Shimazaki, T; Kushiro, A; Watanabe, K; Uchida, K; Yuyama, T; Morotomi, M. Colonization of the stratified squamous epithelium of the nonsecreting area of horse stomach by lactobacilli. *Appl Environ Microbiol*, 2000, 66, 5030-5034.

In: Horses: Biology, Domestication, and Human Interactions
Editor: James E. Leffhalm
ISBN: 978-1-61761-303-6

Chapter 7

HORSE AND HUMAN PERSONALITIES: A DEEP INTERACTION AT WORK

***Georges Gonzalez*[1] *and Paul Sutton*[2]**

[1]INRA-CEFS, Chemin de Borde Rouge, Auzeville, BP 52627, 31326 Castanet-Tolosan Cedex, France.

[2]Cheval Magique – Le Couleil, 09000 Cos, France

ABSTRACT

Scientific publications on animal behavior give increasing attention to personality, regarded in terms of consistent behavioral differences between individuals. And variability in behavioral traits is hypothesized to have important ecological and evolutionary implications. Working from considerations of the origins of domestication, we inquire into the subtle though lasting effect of interacting personalities from both sides of the horse-human relationship.

The early human's capability for horse-keeping developed through a gradual build-up of knowledge on the horse's social organization and basic needs. So at the early stages of horse domestication, humans would have needed to overcome certain cognitive and logistical difficulties as a prerequisite to controlled horse breeding. And this would have required people holding particular skills and possessing particular personality traits.

Since personality traits are the result of natural processes shaped by genetic and ontogenetic determinants, they are eligible as significant factors on both sides of the horse-human relationship. At the horse side, we can suppose that particularly amenable individuals were selectively and repetitively bred at key breeding locations in the emergence of horse lineages. At the human side, we suggest that conditions would have been conducive to development of an in-depth understanding of the horse's behavior and abilities, and that this knowledge would have been harnessed to devise training schemes to make the most of the horse's true capabilities.

INTRODUCTION

Though attempts to examine the influence of horse and human personalities on their relationship from early history to the present time may at first sight appear speculative, it nevertheless appears plausible to hypothesize: i) that the selection schemes engendered by the great equestrian cultures that appeared in historical times would have favored the emergence of horse breeds displaying both mental and physical qualities; and ii) that we might expect at least a partial link between human personality profile and aptitude to change as horses and humans developed together over a period of thousands of years (Fig. 1).

Fig.1. Some horserider relationships history, from left to right: dressage rid, Yakut rider, Numidian warrior and medieval knight.

The earliest evidence for horses being associated with human culture comes from cave paintings 15,000 years old, when horses would have been hunted for meat. Most scholars agree with a datation of 3500 years BPE for the earliest evidence of domestication, in Kazakhstan (Botai Culture – [1]). It seems plausible to assume instances of spontaneous or informal riding at the initial stages of horse domestication, when horses were still human prey. What might have been the personality profiles of these early, "accidental" riders? Bennett ([2] p. 18-19) outlines a possible scenario, involving a horse hunter unexpectedly becoming hay provider to a wounded mare, thus opening the prospect of low-risk access to meat. Managed hunting has been evidenced among pre-Inca Indians with llama and alpaca in Peru [3]. And similar overlaps from forager hunting to gatherer hunting, or even domestication, have been reported for North American Indians with feral horses and by rural peoples during the medieval era in Europe [4]. Many authors underline the difficulties involved in keeping horses in captivity without special facilities, and some authors point out that riding does not imply domestication and might well have preceded it.

EARLY HORSE CULTURES AND EQUESTRIAN CIVILIZATION

Though we obviously cannot pinpoint the personality factors at work in the horse-human relationship thousands of years ago, we can at least look for clues in studies on ancient texts. We find three main socio-economic formations in Antiquity: societies based on sedentary agriculture, tribes of the forest zone, and the nomads of the Eurasian steppe [5]. Nomadism derived from pastoralism, a lifestyle in which economy, culture and lifecycle are heavily dependent on animal husbandry and livestock breeding. From 720 BCE to 300 AD on the north bank of the Black Sea, the domestic horse (ridden using a simple saddle and leather or cloth stirrup) was crucially important to the Cimmerian culture and the Scythian culture that eventually displaced it. Arrival of the horse would have hugely extended the community's capabilities for high-speed, long-duration travel. The importance of agricultural practices would have declined accordingly, under growing demand to increase herd size and acquire pastures. This might well have favored the appearance of a newly predatory attitude to sedentary neighbors: the dark side of nomadism. While pastoral societies were fairly egalitarian, nomadic societies became socially differentiated and, later, socially stratified by genetic lineage [5], as confirmed by finds of lavish equipment and remains of taller horses in excavations of Scythian tombs (Khuran) [6]. Interestingly, in some collective tombs, remains of female riders with their arms have been found [7], suggesting clues as to the status of women in this society.

The first known horsemanship manuals were Hittite Sanskrit tablets relating the Mitanni master horse trainer Kikkuli's teaching on horse chariot training (around 1400 BCE), followed a few centuries later (350 BCE) by the writings of Xenophon, an erudite Greek soldier who detailed the principles of classical dressage, even going so far as to include principles of non-abusive horse training. However this precursor is incidentally reported to have had contemptuous regard for the comfortable saddles of Persian warriors, with more thicknesses of textile than a bed (Vigneron, 1968 in [4]). We can wonder at the training applied to horses to get them to fulfill a useful role in battle: the horse is inescapably a prey species whose fundamental behavior involves rapid flight from real or perceived danger. The contradiction between inherent nature and practical utility appears all the more remarkable when we consider the lightness of the control equipment used in some instances. Ancient descriptions tell of the riding style of the warriors of an ancient Berber kingdom, the Numidians (202 BCE to 46 BCE), riding into battle using a neck strap as sole means of control. And in more recent times the North American Plains Indians bring an example of bareback riding by definite preference, for freedom of movement [4].

By medieval times we see major divergences in riding techniques. One of the most striking is between the standing *jineta* style introduced by the Moors in Spain (711-732), which allowed the horse great freedom of movement, and the seated *brida* style and later hybrid style, *estradiota*, which allowed the rider free-hanging legs. At the time of the Seljukid Dynasty (11th century) three major concomitant equestrian traditions existed, with the Iranians riding stallions, the Turks geldings and the Bedouins mares. It is fascinating to speculate on the possible societal factors underlying these differences; given its crucial importance, a society's approach to the horse will inevitably reflect on its internal structures and workings. A tenuous parallel may be made between the known temperament traits of Arabian breeds, such as excitability (Hausberger et al. 2004, Lloyd et al. 2007) and the atmosphere of traditional events such as the Moroccan *fantasia* and the Asiatic *buzkashi*. We should also bear in mind that that these equestrian artforms are reminiscences of ancestral horse warrior formations (e.g. Numids, Bedouins, Scythians). Again, selection processes will have factored in human and social needs as well as breeding conditions, consistent with the new paradigm of hybrid communities [14].

After the battle of Pavia (1525 AD), a victory of the English lighter cavalry over the heavier French knights, control over the ridden horse further increased with emergence of the first equestrian academies (first in Italy then in France [2]), practicing complex exercises for flexibility and collection. The strong "vertical" image can also be seen as reflecting a sense of power of the horseman over lesser men on foot. Thereafter, academic riding would diverge antagonistically from the more utilitarian military approach, with progress in one perceived as decline in the other [4].

As well as proving invaluable in conquest, horses also lent sheer power in agricultural and other applications. So in the development of horse-human interaction, and thereby in the selective breeding of horse species, we observe a gradual divergence between horse-dominated, often warring, cultures, with their striking horse-based mythologies, and other societies for which the horse would become primarily a valuable work tool. We then observe a period of hyperspecialization, in breeds and breeding regions. At the same time, the horse became omnipresent in human society, and horse-human familiarity developed to an extraordinary degree.

From Utility To Empathy

In the 20th century, human advances in mechanical power inevitably led to a very steep decline in the utilitarian horse. In France, the horse population fell from 3,000,000 in 1913 to just 300,000 to 1989 [4]. Though from a historical perspective we might consider that the perceived individual value of a horse is inversely proportional to its utility, and increases with the time a human spends with it, decline of the work horse in favor of the leisure horse has involved a shift in perception of the animal and its status. And this shift is consistent with research in both this field and that of animal cognition, which challenges the impermeability of the human/animal barrier. Furthermore, perception of these changes is partially under the influence of personality factors.

Historical Background To The Personality Dimension

The notion of rider's personality has surprisingly received little coverage in the literature, despite its relevance to everyday exchanges between humans and horses. Though horsepeople through the ages must have had some realization of how human personality can affect equine behavior, there is little written record of this. We find a glowing historical exception in a manuscript by Portuguese king Dom Duarte around 1434, which mentions the mental preparation required of the rider in order to avoid perturbing the horse by "fragility of character". This text also emphasizes the importance of emotions in the learning process. And it recommends self-knowledge and even puts forward a typology of personalities [8].

Nothing more detailed on the interaction between horse and human personalities is found until recent years, with investigations focusing primarily on rider gender in the relationship with the horse (e.g. Brown 1984 in [9]). We also note investigations on how rider personality appears to alter horse cooperation in emotive horses [10], and findings on differential activity in the cerebral hemispheres during visual identification of a potentially stress-inducing object [11]. Research such as this inevitably tends to elicit speculation along anthropomorphic lines, consistent with a general shift in public attitude to the horse, as people tend to seek parallels between horse and human behavior traits. Paradoxically though, growing interest for the leisure horse often contrasts with actual practices. As riding school schedules fill up, so the time for actual horse-human interaction dwindles, meaning that horse personality factors are shunted into the sidelines. The bond so eagerly sought can only be obtained by time shared by horse and human.

Again looking at horse-human interaction from the angle of human personality, we can assume that a realization of the importance of subjective, personality-driven factors must have existed in traditional approaches to horsemanship, even if only as the result of the biological process of personality development. But though the realization doubtless existed, we might consider that in the Middle Ages, for example, it would have been more difficult than today to outwardly profess an empathy with the horse, then seen primarily and publicly as a tool for war, transport and farming.

THE LEISURE HORSE

We observe two main currents of thought, and of sensitivity, in the shift from work horse to leisure horse:

1 – The first is human-centered in that despite the strong emphasis on animal wellbeing, as regards the minimization of physical and (more recently) mental suffering, the approach remains highly anthropocentric in that the horse is still seen as sort of albeit welcome intruder into the human world. The needs of the human-horse couple are still seen essentially from the human side. Though questions are asked on how the horse experiences things, few questions are asked on the inherent nature of the horse's world. The hierarchical frontier between human and animal remains intact, because the animal is seen through an immutably human frame of reference.

2 – The second is more horse-centered, adopting a perspectivist approach (rooted in the work of earlier philosophers such as Leibniz, Kant, and Nietzsche [12]) that involves apprehending the animal independently of human expectations [13]. This approach is obviously incompatible with a human-centered approach; in both the traditional and the "pet" models, the animal is required to fulfill a function, be it in the form of work or simple presence.

Little work has been done to determine the extent to which the horse-human relationship permeates the horse's world as such [9], though we do have examples of how new aptitudes acquired through this relationship are subsequently applied outside it. This brings us to the new paradigm proposed by the philosopher and ethologist Dominique Lestel under the broader context of human-animal hybrid communities [14].

HYBRID COMMUNITIES AND PERSPECTIVISM

The study of hybrid communities is presented as a new discipline at the interface between human and animal sciences. A hybrid community is one in which human and animal perceive and attribute meaning to the relationship, with each having a capacity to win over the other in order to meet its own expectations. By their very nature, these systems can be highly personalized, involving particular individuals (animal and humans); the notion goes much farther than matching up human and animal groups. We might be tempted to see emergence of the dichotomy between human-centered and horse-centered approaches in the differences reported between boy and girl riders [4]. Boys tend to pursue their equestrian goals forcefully, and often give up when this proves unsuccessful, whereas girls will tend to persevere along more affiliative lines, both on foot and mounted. This fits in with the fact that top competition horses will always have women grooms, presumably better suited to stress-appeasement.

The active role of the horse in the horse-human relationship is more difficult to apprehend, and we know of no research on this. Vinciane Despret's work on Clever Hans [15], a horse with remarkable capacities for observation in response to the expectations of his human experimenters, does however shed light on animal interpretation across the human-animal divide. Recent work shows that the horse's faculty for discrimination is even more acute than we might have expected; in one experiment, horses successfully distinguished between photographs of fraternal and identical twins [16].

The classic approach to evaluation of capacities, involving batteries of tests in an attempt to determine the personality of young horses after training, is behavioristic in that it considers the horse as a stimulus-response system, some properties of which can easily be manipulated (by conditioning and reinforcement) to satisfy the expectations of owners, breeders, trainers, etc. The evaluation results should therefore be regarded with caution.

An alternative and informative approach would consist in examining the training itself rather than the result of training and not only the purpose of the horse-human relationship but the way to reach it [13]. This approach could be illustrated by the phrase "Humans like animals construct their world and so do scientists". The basic meaning can be found in the theoretical framework of Von Huexküll's *umwelts* [17], and in Lestel's notion of hybrid communities [14], in which horse and human learn through interactive contact with each other, each adjusting its presence in the relationship in response to the effects produced on the other. Von Huexküll's *umwelt* notion might be evoked to explain, for example, the poor consensus among riding instructors when asked to evaluate the personality of horses in their own riding school. And in-depth research reports the dramatic change in evaluation quality when evaluators are directly involved in designing the evaluation sheets [18]. So what we actually have is a semiotic dialogue rather than a juxtaposition of stimulus-response systems. Hediger, a zoologist with experience in observing and managing human-animal relations in zoos, was impressed by the ability of certain carers to develop relationships with dangerous species without ever giving the impression of being in danger [13]. This kind of relationship can only really be conceived of in terms of construction of a shared dynamic space.

HORSES AND HUMAN HEALING

Valuable insights into the horse-human relationship come from an examination of the therapeutic effects of contact with horses. Reports point to significant and fairly rapid attenuation in personality disorders of diverse origins among subjects engaged in assisted activities involving horses. The therapeutic effect of contact with horses is not a particularly new discovery; riding sessions were included in therapeutic programs for psychologically damaged German war veterans after the First World War [19]. Many riding instructors express surprise at the passivity of horses involved in therapy, especially with regard to patients liable to “unpredictable” behavior. We might be tempted to claim that horses are sensitive to a sort of behavioral coherence, between mind and body, while realizing that studies to investigate this would probably be highly complex.

More controversially, claims of psychological benefit are made in almost spiritual terms, especially concerning fragile personalities. Proximity to or interaction with horses is seen as capable of revealing deep-lying or often suppressed layers of the human personality. This ties in with the well-known if enigmatic maxim attributed to Hypocrates referring to the curative virtues of the rhythmic motions experienced on horseback: “the outside of a horse is the best possible thing for the inside of a man”. Not all horses, however, are suitable for use in therapeutic applications. If sensory sensitivity is considered a component of equine temperament, complexity rises by virtue of independent variability across the different senses, and the number of possible personalities rises accordingly [20]. Authors also note that it is difficult to objectively determine the suitability of a horse for an equitherapy program on the sole basis of personality or response evaluations, given the influence of other factors such as the smoothness of gait. Another important, and possibly counter-intuitive, fact is that the make-up of a group of riders does not seem to affect the occurrence of stress-induced behaviors among the horses taking part in the sessions, except for groups of children at risk, where an increase in stress-related behavior is observed. To add complexity in this human-horse direction, it has been shown in purely physiological terms that horses are very quick to detect an increase in the rider’s cardiac rate under artificially induced stress conditions [21].

HORSE AND HUMAN PERSONALITIES: AN AFTERWORD

The fact that the horse remains a significant feature in modern human society, despite having outlived its practical utility, suggests either that its function has never been solely utilitarian, or that a very strong cultural dependency has developed: strong enough to survive in the absence of evident practical advantage. In either case, we must assume a psychological link of considerable power. So an examination of horse-human interaction today offers insights into the personality factors that might have permeated the inter-species divide through the long history of this relationship. On the nature of the psychological link, we must begin by postulating a layer of psychology shared across different species. Horse and human psychologies might be considered instantiations of a common mammalian psychology, the mechanisms of which govern phenomena such as the often-reported affinities (or, indeed, discord) between particular riders and horses. It seems plausible to assume that communication between the two species requires a capacity for descending from a species-specific psychological layer into a deeper, shared layer. So what is it that determines a human's capacity to do this? One possible clue comes from Chomskyan linguistics, which holds that language instantiations share coherent settings across a wide range of parameters not immediately perceived to be linked. We might suggest that horses are extremely sensitive to coherence along similar lines. Coherence between leg and hand aids is taught at all riding schools as the most primordial of requirements. We suggest that coherence operates at this and other, deeper, levels. Horses definitely seem able to detect incoherence between apparent and actual approach intentions, so a human wishing to work with wild horses will need either a solid inherent coherence between the two, or an actor's capacity for adaptation. Further evidence comes from the horse's frequently observed tendency to gentleness toward children or disabled people. Might this arise from recognition of a fundamental coherence in the approach made by humans without hidden intentions?

The issue of interaction between horse and human personalities raises the interesting suggestion that instead of developing, as is often assumed, in opposition to other animals, humans have in fact evolved with a characteristic feature of radical interpenetration with "animality", without equivalent in other species. In other words, humans became human by inventing a new pact with "animality", a notion that is curiously corroborated by the comments of art historian Georges Bataille on primitive cave paintings [22]. Rather than attempting an aloof isolation from the animal world, humans pursue the process of humanization by giving full and due place to animal aspects in their representations of themselves, seeing other animals neither as hostile adversaries nor as vestiges of a long-abandoned evolutionary origin, but as an intrinsic part of their make-up.

This article benefited from stimulating discussions with ML. Maublanc.

REFERENCES

[1] Outram, A.K., Stear, N.A., Bendrey, R., Olsen, S., Kasparov, A., Zaibert, V., Thorpe, N., Evershed, R. (2009). The Earliest Horse Harnessing and Milking. Science, 323 (5919), 1332-1335.

[2] Bennett, D. (1998). Conquerors : the roots of New World horsemanship. Los Olivos:Amigo Publications, Inc.

[3] Wheeler, JC., Russel, AJF., Redden H. (1995). Llamas and Alpacas: Pre-conquest breeds and post-conquest hybrids. Journal of Archaeological Science, 22 (6), 833-840.

[4] Digard, JP. (2007). Un histoire du cheval. Arles: Actes Sud.

[5] Khazanov, AM. (1994). Nomads and the outside world. Madison: University of Wisconsin Press.

[6] Subtelny, O. (2000). Ukraine: a history. Toronto: University of Toronto Press.

[7] Adamson, M. (2009). An archaeological analysis of gender roles in ancient non-literate cultures of Eurasia. Master of art thesis, Flinders University of South Australia, Dep. of Archaeology [E-text type]. http://ehlt.flinders.edu.au/archaeology/department/publications/PDF%20Theses/Mike%20Adamson.pdf

[8] Pereira, C. (2009). Parler aux chevaux autrement, approche sémiotique de l'équitation. Paris: Amphora

[9] Mills, DS., McNicolas, J. (2005). The rider-horse relationship. In DS. Mills and SM.McDonnell (Eds.), The domestic horse: the origins, development, and management of its behaviour (161-168). Cambridge : Cambridge University Press.

[10] Visser, EK., Van Reenen, CG., Blokhuis, MZ., Morgan, EKM., Hassmen, P., Rundgren, TMM., Blokhuis, HJ. (2008). Does Horse Temperament Influence Horse–Rider Cooperation? Journal of Applied Animal Welfare Science, 11(3), 267 – 284.

[11] De Boyer Des Roches, A., Richard-Yris, MA ., Henry, S., Ezzaouia, M., Hausberger, M. (2008). Laterality and emotions: Visual laterality in the domestic horse (Equus caballus) differs with objects' emotional value. Physiology & Behavior, 94 (3), 487-490.

[12] Giere, RN. (2006). Scientific perspectivism. Chicago: University of Chicago Press.

[13] Noske, BM. (2005). Horse images and the human self-image in equine research. In de Jonge, FH., van den Bos, R.. (Eds.), The human-animal relationship: forever and a day (38-50). Assen : Uitgeverij Van Gorcum.

[14] Lestel, D., Brunois, F., Gaunet, F. (2006). Etho-ethnology and ethno-ethology. Social Science Information, 45 (2), 155-177.

[15] Despret, V. (2004). Hans, le cheval qui savait compter. Paris : Le Seuil/ Les Empêcheurs de penser en rond.

[16] Stone, S. (2009). Human facial discrimination in horses: can they tell us apart? Animal Cognition, 13, 51-61.

[17] Noske, B. (1997). Beyond boundaries: Humans and animals. Montreal: Black Rose Books.

[18] Grajfoner, D. (2007). The role of observer individual differences in personality assessments of the domesticated horse: A novel application of Kelly's Repertory Grid Technique. PhD Thesis in Psychology. The University of Edinburg.

[19] Scott, N. (2005). Special needs, special horses: a guide to the benefits of therapeutic riding. Denton: University of North Texas Press.

[20] Lansade, L., Pichard, G., Leconte, M. (2008). Sensory sensitivities: Components of a horse's temperament dimension. Applied Animal Behaviour Science, 114 (3), 534-553.

[21] Keeling LJ., Jonare L. and Lanneborn L. 2009. Investigating horse–human interactions: The effect of a nervous human. The Veterinary Journal, 181, 70-71.

[22] Bataille, G. (1955). Lascaux or the Birth of Art. Geneva: Skira.

[23] Hausberger, M., Bruderer, C., Le Scolan, N. & Pierre, J.S. (2004). Interplay Between Environmental and Genetic Factors in Temperament/Personality Traits in Horses (Equus caballus). Journal of Comparative Psychology, 118(4), 434-446.

[24] Lloyd, A.S., Martina, J.E., Bornett-Gaucia, H.L.I. & Wilkinson, R.G. (2007). Evaluation of a novel method of horse personality assessment: Rater-agreement and links to behaviour. *Applied Animal Behaviour Science, 105(1-3)*, 205-222.

In: Horses: Biology, Domestication, and Human Interactions ISBN: 978-1-61761-303-6
Editor: James E. Leffhalm

Chapter 8

ANALYSIS OF 16 MICROSATELLITES MARKERS AND PARENTAGE TESTING IN ITALIAN STANDARDBRED HORSE

M. Dobosz[1*], C. Bocci, M. Bonuglia, C. Grasso, A. Russo and P. De Iuliis
DPT of Clinical and Experimental Medicine, Forensic Medicine Section, University of Perugia, via del Giochetto – 06126 Perugia – Italy

SUMMARY

Accurate determination of relatedness is of great importance in the horse breeding industry. In the past decade, the DNA forensic field was primarily focused on human genome typing, for identification and parentage determination purposes. The discovery of polymorphisms in microsatellites loci (STRs) and the introduction of polymerase chain reaction technique has lead to the establishment of an extremely powerful method for identification and for parentage analysis also in animals. Microsatellites have been in fact identified in all eukaryotic species studied so far and modern tools of molecular biology provided means for a fast and accurate way for animal genotyping. Currently, the majority of horse breed registries adopted DNA technology in order to verify pedigree records and to solve queries of parentage.

In this study we analyzed 16 microsatellites markers in a population of 14,110 Italian Standardbred foals. Among these, 358 unrelated individuals were selected in order to evaluate the feasibility of the STRs panel to verify parentage and individual identification. Number of alleles, heterozygosities, polymorphic information content (PIC) and exclusion probabilities of each system were calculated. The mean polymorphic information content (PIC) is 0.67 and the exclusion probability (PE) of a false parent of the system is 0.99998. Mutational events observed are at an acceptable level for parentage testing.

* Corresponding author: U.N.I.R.E.LAB s.r.l., Veterinary Forensic Genetics Laboratory,Via Campobello 7/c - 00040 Pomezia (RM) – Italy. E-mail: marinadobosz@libero.it, tel./fax +39 75 5857340

Keywords: Italian Standardbred horse, microsatellites, population genetics, parentage verification

INTRODUCTION

Historically, the Italian Standardbred breed was formed by mating mares of different breeds which proved to have some qualities in harness racing, with foreign stallions, mainly Thoroughbred, followed by Russian, Norman and American Standardbred stallions. Several studies based on the distribution of mitocondrial DNA haplotypes in modern horses suggested that horse domestication probably involved distinct populations. The initially domesticated horses spread out and incorporated wild mares forming different mtDNA clusters. Horse population breeds in Italy are the result of multiple origins since they retain very ancient mitochondrial diversity (Cozzi et al. 2004). Data on microsatellites markers confirmed that the formation of the Standardbred consisted of several breeds, including the Thoroughbred (Bjurnstad and Rued 2001).

The Italian Standardbred studbook was founded in 1896 and includes offspring born in Italy from mares and stallions registered in the book. To improve peculiar characteristics of the Italian breed, only a few horses included in foreign studbooks can be admitted as mares and stallions. To protect studbook integrity, horse breed registries worldwide rely on genetic testing (Tozaki et al. 2001). Microsatellites are highly polymorphic in humans and widely used in forensic analysis (Gill et al. 1995; Goldstein and Schlotterer 1999). DNA technology was simultaneously developed also for animals, plants and microorganisms. Modern tools of molecular biology provided means for an accurate way for animal genotyping. Recently, a large number of microsatellites have been identified from the equine genome and many protocols were performed to analyze pedigrees and verify parentage of different breeds (Dimsoski 2003; Lee and Cho 2006;Tozaki et al. 2003). Microsatellites, in particular the (CA/TG)n repeats, which were constructed with tandemly repeated DNA sequences consisting of poly(dC-dA)/(dG-dT) repeats, have been found to be common in all eukaryotic genomes, occurring once for every 30–60 kb (Weber and May 1989). As microsatellites are also frequently polymorphic and evolutionarily conserved in the eukaryotic genome, they provide useful markers for comparative studies of genetic variation, parentage testing, and studies of gene flow in many horse populations (Cunningham et al. 2001). However, only few data on microsatellites genetic variability in Italian Standardbred breed have been reported (Bruzzone et al. 2003).

At present, parentage testing of horses is carried out using a minimum panel of 12 microsatellites loci recommended by the International Society for Animal Genetics. However, in some cases, this panel is not useful to exclude a relationship. In Thoroughbred horses, for example, one dam can mate with multiple sires in one breeding season, and the candidate sires can be closely related. In these cases, to exclude a false sire, more markers with a higher PE value are needed (Tozaki et al. 2001).

In the present study we performed a routine DNA typing with the following microsatellites loci: AHT4, AHT5, ASB2, ASB17, ASB23, CA425, HTG4, HTG6, HTG7, HTG10, HMS2, HMS3, HMS6, HMS7, VHL20, HMS1 (Table 1). These markers include nine loci recommended by the Equine Genetics and Thoroughbred Parentage testing

Standardization Committee of the International Society for Animal Genetics and eight additional loci commonly used for horse parentage and identification testing (Dimsoski 2003; Lear et al. 1999). These markers, initially characterized in the Thoroughbred horse, have been used for genetic typing of different horse breeds (Azor et al. 2007; Giacomoni et al. 2008).

Table 1. Microsatellite markers used in this study

Marker	Primer sequences 5'- 3'	Allele size (bp)	Reference
AHT4	AACCGCCTGAGCAAGGAAGT GCTCCCAGAGAGTTTACCCT	138-170	Binns et al.,1995
AHT5	ACGGACACATCCCTGCCTGC GCAGGCTAAGGGGGCTCAGC	128-152	Binns et al.,1995
ASB2	CCACTAAGTGTCGTTTCAGAAGG CACAACTGAGTTCTCTGATAGG	222-256	Breen et al. (1997)
ASB17	GAGGGCGGTACCTTTGTACC ACCAGTCAGGATCTCCACCG	89-131	Breen et al. (1997)
ASB23	GCAAGGATGAAGAGGGCAGC CTGGTGGGTTAGATGAGAAGTC	176-212	Irvin et al. (1998)
CA425	AGCTGCCTCGTTAATTCA CTCATGTCCGCTTGTCTC	230-250	Eggleston-Stott et al. (1997)
HMS1	CATCACTCTTCATGTCTGCTTGG TTGACATAAATGCTTATCCTATGGC	166-178	Guerin et al. (1994)
HMS2	TTGACATAAATGCTTATCCTATGGC ACGGTGGCAACTGCCAAGGAAG	220-236	Guerin et al. (1994)
HMS3	CCAACTCTTTGTCACATAACAAGA CCATCCTCACTTTTTCACTTTGTT	150-174	Guerin et al. (1994)
HMS6	GAAGCTGCCAGTATTCAACCATTG CTCCATCTTGTGAAGTGTAACTCA	153-171	Guerin et al. (1994)
HMS7	CAGGAAACTCATGTTGATACCATC TGTTGTTGAAACATACCTTGACTGT	167-189	Guerin et al. (1994)
HTG4	CTATCTCAGTCTTGATTGCAGGAC CTCCCTCCCTCCCTCTGTTCTC	127-141	Ellegren et al. (1992)
HTG6	TGGAAATAAGGTTAGCAGGGATGC CCTGCTTGGAGGCTGTGATAAGAT	82-106	Ellegren et al. (1992)
HTG7	CCTGAAGCAGAACATCCCTCCTTG ATAAAGTGTCTGGGCAGAGCTGCT	114-132	Marklund et al. (1994)
HTG10	CAATTCCCGCCCCACCCCCGGCA TTTTTATTCTGATCTGTCACATTT	89-171	Marklund et al. (1994)
VHL20	CAAGTCCTCTTACTTGAAGACTAG AACTCAGGGAGAATCTTCCTGAG	89-107	van Haeringen et al. (1994)

MATERIAL AND METHODS

Sampling and DNA Extraction

Two automatic workstations, TECAN GENESIS RSP 150 and HAMILTON MICROLAB® STAR, have been set up to analyze DNA from biological samples (FTA®, hair roots, blood, urine, semen). In this study genomic DNA was extracted from blood spotted into FTA® filter papers (Whatman Bioscience, Cambridge, UK). FTA® card are impregnated with a patented chemical formula that lyses cell membranes and denatures proteins upon contact and nucleic acid are physically entrapped, immobilized and stabilized for storage at room temperature (Tack et al., a2007). This technology provides a method designed to simplify collection, shipment, archiving and purification of nucleic acids. From each sample, one FTA® disk of 1.2 mm in diameter was punched into 96-well microplates, transferred on the robotic workstations. One negative and one positive control are used in each plate. All samples are processed according the manufacturer's protocol (FTA® Protocols, Whatman, 2002). After DNA extraction, the robotic workstations prepare the PCR master mix and dispense it with volume's reduction, saving time and costs. Best results can be reached with 25% reaction recommended volume (3,5 μl). PCR mix reaction can be added directly onto FTA punch. A 1.2 mm disc contains about 5-20 ng DNA.

Microsatellites and Population Genetics Analysis

Microsatellites were co-amplified with Equine Genotypes™ Panel 1.1 kit (Finnzymes Diagnostics, Finland), according following conditions: initial denaturation for 3 min at 98°, followed by 30 cicles at 98°C for 15 sec, 60°C for 75 sec, 72°C for 30 sec. An extension step of 72°C for 5 min was added at the final cycle. Experiments were performed in GeneAmp PCR System 9700 (Applid Biosystem, USA). After the amplification process, PCR products were separated by allele capillary electrophoresis on ABI PRISM 3100 Genetic Analyzer (Applied Biosystems, USA). Size analysis of DNA fragments was performed with Genotyper Software Ver. 3.7 (Applied Biosystems, USA). The internal standard Genescan-LIZ 500 was used for sizing alleles. In addition, samples of known genotypes from ISAG comparison tests were used as reference to standardize sizes.

Experiments were performed according the ISAG guidelines.

The entire process, from sample registration to final diagnosis, was managed by SQL*LIMS®4 software (Applied Biosystem).The program is a powerful laboratory information management system able to interface with software and laboratory instruments (Garofano et al. 2004).

RESULTS

Heterozygosities and Number of Alleles

Allele frequencies, heterozygosity, polymorphic information content (PIC) for each locus were computed using the DNA VIEW software program (Brenner 1989-2007), as were the probabilities of exclusion. The locus LEX03 was not included in the calculations because, due to it's location on the X chromosome, it hasn't paternal allele.

The DNA-VIEW software is an acknowledged worldwide leader and standard statistical program for DNA identification. Leading paternity and crime laboratories have been using the software since 1988. Recently a new version (v. 28.55) has been implemented with a new *family* computation which evaluates whether both mare and stallion are the correct parents of a given foal. The new *Calculate family trio* option in the Paternity case menu is a convenience which tests for both paternity and maternity. It performs the usual paternity calculation and then does a calculation for the mother as well. This feature is useful when dealing with livestock such as horses, where maternity is less certain than is usual with people. A DNA.VIEW/LIMS interface was performed to allow the program to be driven as a computational engine by a LIMS system, normally without user interaction with DNA.VIEW.

Table 2. Heterozygosity (H_O/H_E), polymorphic information content (PIC) and exclusion probability (PE) of each marker

Marker	Chromosome location	k	H_O	H_E	PIC	PE
A H T4	24	6	0.737	0.704	0.668	0,489
A H T5	6	6	0.737	0.737	0.693	0,489
A S B 17	2	12	0.813	0.800	0.775	0,618
A S B2	15	8	0.816	0.831	0.808	0,628
A S B 23	3	7	0.763	0.729	0.687	0,529
C A 4 25	28	8	0.765	0.788	0.756	0,539
H M S1	15	7	0.735	0.703	0.663	0,486
H M S2	10	9	0.844	0.796	0.768	0,684
H M S3	9	7	0.746	0.725	0.690	0,505
H M S6	4	5	0.712	0.750	0.708	0,450
H M S 7	1	7	0.623	0.659	0.596	0,322
H T G 10	21	9	0.587	0.598	0.568	0,278
H T G 4	9	5	0.788	0.765	0.724	0,579
H T G6	15	5	0.567	0.576	0.490	0,256
H T G 7	4	4	0.494	0.483	0.450	0,185
V H L 20	30	8	0.782	0.761	0.720	0,568
Mean		7.06	0,713		0.672	0,99998*

K: numer of alleles; H_O: observed heterozygosity; H_E: expected heterozygosity.

* Total exclusion probability

Table 3. Allele frequencies of microsatellites analyzed in Italian Standardbred horse

Marker	F	G	H	I	J	K	L	M	N	O	P	Q	R	S	T	U
A H T4			0.170	0.123	0.092	0.007				0.476	0.131					
A H T5					0.376	0.214	0.008	0.254	0.098	0.050						
A S B 17	0.145	0.059			0.001	0.001	0.001	0.050	0.176	0.112		0.004	0.346	0.101		0.003
A S B2				0.022		0.203		0.233	0.108	0.196	0.042	0.060	0.137			
A S B 23				0.109	0.404	0.284	0.120							0.028	0.031	0.025
C A 4 25	0.039			0.003	0.279	0.112	0.005	0.124	0.287	0.149						
H M S1				0.130	0.465	0.169	0.011	0.187	0.029		0.008					
H M S2			0.055	0.003	0.028	0.116	0.334	0.008		0.205	0.162		0.089			
H M S3				0.159				0.026	0.169	0.003	0.448	0.072	0.122			
H M S6						0.078	0.116	0.184		0.359	0.263					
H M S 7					0.098	0.008	0.429	0.056	0.380	0.028		0.001				
H T G 10				0.603		0.006	0.004	0.063	0.064	0.130	0.011	0.003	0.116			
H T G 4						0.259	0.140	0.292	0.244	0.064						
H T G6		0.379		0.010	0.524				0.001	0.087						
H T G 7						0.070		0.127	0.106	0.697						
V H L 20				0.019			0.217	0.275	0.302	0.015	0.007	0.006	0.159			

Data obtained from statistical analysis revealed that microsatellites are highly polymorphic in Italian Standardbred horse (table 2 and 3). A total of 113 alleles was found, and the number varies from 4 for HTG7 locus to 12 for ASB17 locus. The average number of alleles is 7,06.

The observed and expected heterozygosity ranged from 0.494 to 0.844 and from 0.483 to 0.831 respectively. Not significant deviations from Hardy-Weinberg expectation were found. PIC value ranged from 0.450 to 0.808 with a mean value of 0.672. The total probability value to exclude a false parent is 0,99998.

Parentage Verification

Alphabetical allele codes for all loci are identical to the assignment on 2000 ISAG horse comparison test. Of the 14,110 foals typed for parentage testing, 13,998 (92,2%) were qualified by the compatibility of 16 microsatellites markers according to Mendelian fashion. 105 foals were excluded with at least two genetic incompatibilities with one/both parents. In the remaining 7 cases the offspring revealed genetic incompatibility with one parent only at a single locus. 13 additional STRs (Tozaki et al. 2001) failed to identify other incompatibilities, for a total of 30 loci tested. These cases were considered as mutational events, probably due to a stepwise mutation model, the simplest model of microsatellites mutation: during replication of the microsatellite region the strands may be displaced and then realign incorrectly, leading to insert or delete 1 repeat unit (Otha and Kimura 1973; Whittaker et al. 2003). Such mutations were observed at loci ASB2, ASB17, ASB23, HMS6, HMS7, VHL20; based on the 716 meiotic events considered in this study, they provided a mutation estimate of $2.8x10^{-3}$ / generation for VHL20 locus and of $1.4x10^{-3}$/ generation for other loci.

Discussion

Analysis of genetic diversity is today largely based on the use of the polymerase chain reaction (PCR). Already, a considerable number of PCR-based sequences are known, in human and other genomes, and the number is steadely increasing. To be useful for the identification of individuals, the markers should fulfill certain criteria: pattern of inheritance well established, high degree of polymorphism, independent inheritance from the other markers used, population data of alleles and genotype frequency established, low mutation rate. Microsatellites loci are particularly suitable for identification purposes, since they are highly polymorphic and they exhibit a high degree of heterozygosity. Moreover, the technique for the analysis of these informative markers is simple, rapid and reproducible. Multiple STRs have been validated in genetic mapping, linkage analysis and forensic identification in humans and animals. In horse genotyping, highly discriminatory panels were designed. Nevertheless, only few genetic studies have been performed on Standardbred Horse breeds with microsatellites markers. Bruzzone et al. (2003) analyzed 12 microsatellites of 40 Italian Standardbreds to compare them to other domestic horse breeds. Azor et al (2007) used 16 microsatellites to genotype 40 Spanish Standardbreds. Bjornstad et al. (1999, 2001a,b) analyzed genetic variability of 40 Norwegian Standardbreds and 44 Colblooded

Standardbreds with 10 and 26 microsatellites respectively. With this research we investigated 16 microsatellites markers in Italian Standardbreds, with the aim to improve information on STRs variability in Italian horse breeds. Data on microsatellites markers reported in the literature for other horse breeds indicate an observed heterozygosity mostly ranging from 0.610 to 0.75 for (Achmann et al. 2004; Behl et al. 2006; Galov et al. 2005; Giacomoni et al. 2008; Loùis et al., 2007; Tozaki et al. 2001, 2003), although these values are not completely comparable with those detected by us because of differences in the microsatellite sets used. In parentage testing, the usefulness of any co-dominant marker is defined as the probability of it making an exclusion. The breed analyzed in the present study showed high microsatellites variability **and t**he probability of the 16 microsatellites panel to exclude a false parent is 0.99998. Mutational events observed in 358 parent-offspring triplets provided rate estimates of 2.8×10^{-3} / generation for VHL20 locus and 1.4×10^{-3}/ generation for other loci (ASB2, ASB17, ASB23, HMS6, HMS7). These data are in agreement with mutation rates of STRs markers used in human genetic testing (Gjertson et al. 2007), suggesting that the 16 microsatellites panel is greatly useful for parentage verification in the Italian Standardbred horse.

ACKNOWLEDGMENTS

The authors thank Charles Brenner for his help in using the DNA.VIEW program, Federico De Palma, Andrea Rossi and Paolo Zanini for their technical abilities in implementing the DNA.VIEW/LIMS interface.

REFERENCES

Achmann, R., Curik, I., Dovc, P., Kavar, T., Bodo, I., Habe, F., Marti, E., Solkner, J. & Brem, G. (2004). Microsatellites diversity, population subdivision and gene flow in the Lipizzan Horse. *Animal Genetics, 35*, 285-292.

Behl, R., Behl, J., Gupta, N., Gupta, S. C., Ahlawat, S. P. S., Ragnekar, M. & Ahmed, Z. (2006). Genetic characterization of Zanskari breed of horse. *Journal of Genetics, 85*, 199-203.

Binns, M. M., Holmes, N. G., Holliman, A. & Scott, A. M. (1995). The identification of polymorphic microsatellite loci in the horse and their use in thoroughbred parentage testing. *British Veterinary Journal, 151(1)*, 9-15.

Bjurnstad, G. & Rued, K. H. (2001) Breed demarcation and potenzial for breed allocation of horses assessed by microsatellites markers. *Animal Genetics, 32*, 59-65.

Bjurnstad, G., Gunby, E. & Roed, K. H. (2000a). Genetic structure of Norwegian horse breeds. *Journal of Animal Breeding and Genetics, 117*, 307-317.

Bjurnstad, G., Midthjell, L. & Rued, K. H. (2000b). Characterization of ten equine dinucleotide microsatellites loci: NVHEQ21, NVHEQ54, NVHEQ67, NVHEQ70, NVHEQ75, NVHEQ77, NVHEQ79, NVHEQ81, NVHEQ82 and NVHEQ83. *Animal Genetics, 31*, 78-79.

Bowling, A. T., Eggleston-Stott, M. L., Byrns, G., Clark, R. S., Dileanis, S. & Wictum, E. (1997). Validation of microsatellite markers for routine horse parentage testing. *Animal Genetics, 28*, 247-252.

Brenner, C. H. (1989-2007). DNA User's Manual.

Bruzzone, A., Iamartino, D., Blasi, M. & Pilla, F. (2003). The Pentro Horse: genetic characterization by microsatellites markers. *Italian Journal of Animal Science, 2*, 223-203.

Budowle, B., Garofano, P., Hellman, A., Ketchum, M., Sree, S., Kanthaswamy, S., Parson, W., van Haeringen, W., Fain, S. & Broad, T. (2005). Recommendations for animal DNA. *International Journal of Legal Medicine, 119(5)*, 295-302.

Cozzi, M. C., Strillacci, M. G., Valiati, P., Bighignoli, B., Cancedda, M. & Zanotti, M. (2004). Mitochondrial D-loop sequence variation among Italian horse breeds. *Genetics Selection Evolution, 36*, 663-672.

Cunningham, E. P., Dooley, J. J., Splan, R. K. & Bradley, D. G. (2001). Microsatellite diversity, pedigree relatedness and the contributions of founder lineages to thoroughbred horses. *Animal Genetics, 32*, 360-364.

Dimsoski, P. (2003). Development of a 17-plex Microsatellite Polymerase Chain Reaction Kit for Genotyping Horses. *Croatian Medical Journal, 44(3)*, 332-335.

Eggleston-Stott, M. L., DelValle, A., Bautista, M., Dileanis, S., Wictum, E. & Bowling, A. T. (1997). Nine equine dinucleotide repeats at microsatellites loci UCDEQ136, UCDEQ405, UCDEQ412, UCDEQ425, UCDEQ437, UCDEQ467, UCDEQ487, UCDEQ502 and UCDEQ505. *Animal Genetics, 28*, 370-371.

Ellegren, H., Johansson, M., Sandberg, K. & Andersson, L. (1992). Cloning of highly polymorphic microsatellites in the horse. *Animal Genetics, 23*, 133-142.

Galov, A., Byrne, K., Duras-Gomercic, M., Gomercic, T., Nushol, Z., Vincek, Z., Kocijan, D. I., Tadić, Z., Benković, V., Bašić, I. & Funk, S. M. (2005). Effectiveness of nine polymorphic microsatellites markers in parentage testing in Posavina, Croatian Coldblood and Lipizzaner horse breeds in Croatia. *Livestock Production Science, 93*, 277-282.

Garofano, P., Veneroni, R., Molinari, S., Bonuglia, M., De Iuliis, P. & Montana, M. (2004). High throughput dna isolation and STRs analysis for animal identification: forensic casework and DNA database applications in animal genetics. *Proceedings of the 15th International Symposium on Human Identification*-Phoenix, Oct 4-7.

Giacomoni, E. H., Fernandez-Stolz, G. P. & Freitas, T. R. O. (2008). Genetic diversity in the Pantaneiro horse breed assessed using microsatellites DNA markers. *Genetics and Molecular Research, 7*, 1, 261-270.

Gill, P., Kimpton, C. P., Urquhart, A., Oldroyd, N., Millican, E. S., Watson, S. K. & Downes, T. J. (1995). Automated short tandem repeat (STR) analysis in forensic casework - a strategy for the future. *Electrophoresis, 16*, 9, 1543-1552.

Gjertson, D.W., Brenner, C. H., Baur, M. P., Carracedo, A., Guidet, F., Luque, J. A., Lessig, R., Mayr, W. R., Pascali, V. L., Prinz, M., Schneider, M. P. & Morling, N. (2007). ISFG: Recommendations on biostatistics in paternity testing. *Forensic Science International: Genetics, 1*, 223-231.

Goldstein, D. B. & Schlotterer, C. (1999). *Microsatellites: Evolution and Applications,* Oxford University Press, Oxford, UK.

Guérin, G., Bertaud, M. & Amigues, Y. (1994). Characterization of seven new horse microsatellites: HMS1, HMS2, HMS3, HMS5, HMS6, HMS7 and HMS8. *Animal Genetics, 25*, 62.

Irvin, Z., Giffard, J., Brandon, R., Breen, M. & Bell, K. (1998). Equine dinucleotide repeat polymorphisms at loci ASB21, 23, 25 and 37-43. *Animal Genetics, 29*, 67.

Lear, T. L., Brandon, R. & Bell, K. (1999). Physical mapping of ten equine dinucleotide repeat microsatellites. *Animal Genetics, 30*, 3, 235.

Lee, S. & Cho, G. (2006). *Parentage testing* of Thoroughbred *horse* in Korea using microsatellite DNA typing. *Journal of Veterinary Science, 7*, 1, 63-67.

Lùis, C., Juras, R., Oom, M. M. & Cothran, E. G. (2007). Genetic diversity and relationship of Portuguese and other horse breeds based on protein and microsatellites loci variation. *Animal Genetics, 38*, 20-27.

Marklund, S., Ellegren, H., Eriksson, S., Sandberg, K. & Andersson, L. (1994). Parentage testing and linkage analysis in the horse using a set of highly polymorphic microsatellites. *Animal Genetics, 25*,19-23.

Ohta, T. & Kimura, M. (1973). The model of mutation appropriate to calculate the number of electrophoretically detectable alleles in a genetic population. *Genetic Research, 22*, 201-204.

Tack, L. C., aThomas, M. S. b. & Reich, K. (2007). Automated Forensic DNA Purification Optimized for FTA Card Punches and Identifiler STR-based PCR Analysis. *Clinics in Laboratory Medicine*, *27, 1*, 183-191.

Tozaki, T., Kakoi, H., Mashima, S., Hirota, K. I., Hasegawa, T., Ishida, N., Miura, N., Choi-Miura, N. H. & Tomita, M. (2001). Population study and validation of paternity testing for Thoroughbred horses by 15 microsatellite loci. *Journal of Veterinary Medical Science, 63*, 1191-1197.

Tozaki, T.,Takezaki, N., Hasegawa, T., Ishida, N., Kurosawa, M., Tomita, M., Saitou, N. & Mukoyama, H. (2003). Microsatellite Variation in Japanese and Asian Horses and Their Phylogenetic Relationship Using a European Horse Outgroup. *Journal of Heredity, 94*, 374-380.

van Haeringen, H., Bowling, A. T., Stott, M. L., Lenstra, J. A. & Zwaagstra, K. A. (1994). A highly polymorphic horse microsatellite locus: VHL20. *Animal Genetics, 25*, 207.

Weber, J. L. & May, P. E. (1989). Abundant class of human DNA polymorphisms which can be typed using the polymerase chain reaction. *American Journal of Human Genetics, 44*, 388-396.

Weber, J. L. & Wong, C. (1993). Mutation of human short tandem repeats. *Human Molecular Genetics, 2*,1123-1128.

Whittaker, J. C., Harbord, R. M., Boxall, N., Mackay, I., Dawson, G. & Richard, M. Sibly, M. (2003). Likelihood-Based Estimation of Microsatellite Mutation Rates. *Genetics* 164, 781-787.

In: Horses: Biology, Domestication, and Human Interactions
Editor: James E. Leffhalm
ISBN: 978-1-61761-303-6

Chapter 9

EQUINE MICROSATELLITE ANALYSIS OF CHALLENGING SAMPLES

C. Grasso,*[1] M. Dobosz[1], C. Bocci[2], M. Bonuglia[2], A. Russo[2] and P. De Iuliis[2]

[1]DPT of Clinical and Sperimental Medicine, Forensic Medicine Section, University of Perugia, via del Giochetto – 06126 Perugia – Italy
[2]UNIRELAB s.r.l., Veterinary Forensic Genetics Laboratory, via Campobello 7/c - 00040 Pomezia (RM) – Italy

INTRODUCTION

Horse identification and parentage control are established in all countries where equestrian sports are developed. As in human forensics, the horse identity can be confirmed by DNA typing.

Blood, saliva, hair roots and tissue fragments provide a reliable source of DNA, but, when it is impossible to recover those biological samples, DNA can be extract from "challenging" samples, such as bone, tooth and urine samples.

MICROSATELLITE ANALYSIS

At present, microsatellites loci (STRs) are the best markers for identification purposes in horse: in fact, compared to the traditional blood markers, STRs have the advantage of being numerous, highly polymorphic and well spread in the whole genome.

Moreover microsatellites are transmitted according to Mendelian inheritance, so that they are useful in the paternity testing of horses (Marklund et al., 1994; Binns et al., 1995;

* Corresponding author: c.grasso@unirelab.it

Usha et al., 1995; Bowling et al., 1997; Tozaki et al., 2001). More than thirty of these markers were adopted by ISAG/FAO Working group for the study of the genetic diversity of horse breeds (Hoffmann et al., 2004).

In this study, the following microsatellite loci were analysed: AHT4, AHT5, ASB2, ASB17, ASB23, CA425, HMS1, HMS2, HMS3, HMS6, HMS7, HTG10, HTG4, HTG6, HTG7, LEX03, VHL20. These markers include nine loci recommended by the "Equine Genetics and Thoroughbred Parentage testing Standardization Committee of the International Society for Animal Genetics" and eight additional loci commonly used for horse parentage and identification testing (Dimsoski 2003; Lear et al., 1999; Azor et al., 2007; Giacomoni et al. , 2008).

Bone and Tooth Samples

Genetic typing can be performed in cases of extreme degradation of equine remains and from fossils where bones or teeth are the unique material available.

Bone and tooth sample provide a physical barriers to protect DNA from action of degrading enzymes; for example, the hydroxyapatite mineral, which is a major component of these hard tissues, stabilizes the DNA.

Osteocytes are the most common nucleated cells in the bone matrix, while in the teeth, the pulp cavity provide a source of nucleated cells.

Materials and Methods

DNA Isolation

Bones surfaces were cleaned and rinsed with 5% bleach to remove any contamination and any possible traces of DNA degraders or PCR inhibitors, such as haemoglobin.

The bones were washed with de-ionized water and left to air dry for 24h.
Samples were pulverized into fine powder using a Freezer/mill, where 500 mg of bone powder was removed, decalcified in 8 ml 0.5 M EDTA (pH 8.0) with continuous vertical rotation for 48 h. During that period, the EDTA solution was replace after 24 h.

After centrifugation, the supernatant was discarded. The resulting pellet was digested in 8 ml lysis buffer (7.18 ml 0.5 M EDTA, pH 8.0, 800μl 10% SDS, 20 μl 20 mg/ml Proteinase K) and incubated, with shaking, at 48°C for 24h (Davoren et al., 2007; Loreille et al., 2006; Kalmár et al., 2000; Rohland et al.,2007).

For the teeth, the external surface was washed with distilled water and then crushed into small fragments. Only the pieces of root body from the fragments were digested in the lysis buffer, as these had the highest DNA yield (Gaytmenn, R, et al., 2003).

After centrifugation, the supernatant was added to 6 ml phenol/chloroform/isoamyl alcohol mixture (25:24:1), mixed and centrifuged for 3 min; the aqueous phase was added to 6 ml chloroform/isoamyl alcohol (24:1), mixed and centrifuged for 3 min. The aqueous phase was then concentrated with a Amicon Ultra-30 Centrifugal Filter device and centrifuged at 14,000 rpm until 200 μl remained.

This organic method yields amounts of DNA with higher levels of PCR inhibition: it could be useful performing an additional extraction protocol using PrepFiler™Forensic DNA Extraction Kit (Applied Biosystems) (Brevnov M.G., et al., 2009) and purifying by FAM PCR&GEL Purification kit (FAMBiotech s.r.l.).

To avoid potential contamination, extraction blanks containing everything except bone powders or tooth sample were carried through the entire process. The extraction blanks were negative for STRs amplification.

The DNA yield was evaluated by spectrophotometric analysis at 260 nm, using a NanoDrop ND-1000 Spectrophotometer (NanoDrop Technologies, Inc., USA).

DNA concentration ranged from 0.2 to 3 ng/µl. The quality of DNA was evaluated by estimating the ratio of absorbance at 260 nm and 280 nm, with an average value of 1.60.

Microsatellites were amplified with Equine Genotypes™ Panel 1.1 kit (Finnzymes Diagnostics, Finland), in accordance with the manufacturer's protocol. PCR was performed in GeneAmp PCR System 9700 (Applied Biosystem, USA).

PCR products were separated by capillary electrophoresis on ABI PRISM 3130*xl* Genetic Analyzer (Applied Biosystems, USA). Size analysis of DNA fragments was performed with GeneMapper Software v3.0 (Applied Biosystems, USA).

The internal standard Genescan-LIZ 500 was used for sizing alleles. In addition, samples of known genotypes from ISAG comparison tests were used as reference to standardize the alleles.

Experiments were performed according to the ISAG guidelines (Budowle et al., 2005).

Results And Discussion

DNA profiles obtained show a variable quality depending by sample: teeth DNA profiles are better than the bone's one; moreover, it seems reasonable to suppose that the characteristics of the bone can influence the yield of DNA (Milos et al.,2007).

Successful DNA typing does not depend on the amount of sample, but depends on the absence of PCR inhibitors. Phenol is known as a strong PCR inhibitor, but both the additional extraction method and purification kit gave successful amplification of 9 or 16 STR loci.

We have obtained good results from samples that failed previously. These results demonstrate that this particular purification procedure increases the success rate of PCR amplification and STR analysis. Bone sample DNA profile obtained with different extraction procedures is shown in **Fig.1.**

We have compared these genetic results to the DNA profiles obtained from hair or blood of the same horses, stored in U.N.I.R.E. forensic database. The comparison shows that all the DNA analysed are identical to the DNA profile of the reference samples.

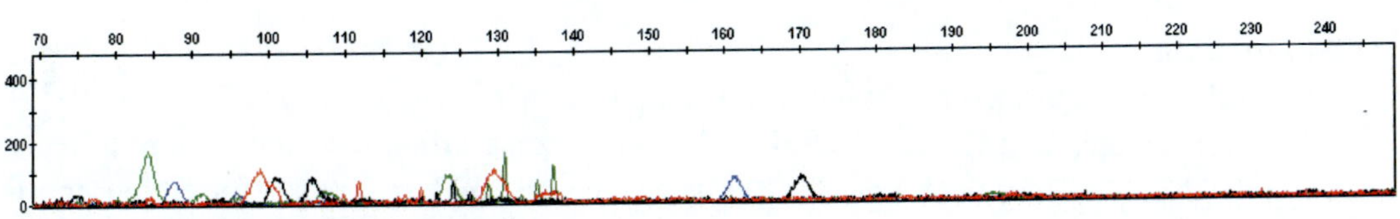

a) phenol/chloroform DNA extraction

Figure 1 (Continued)

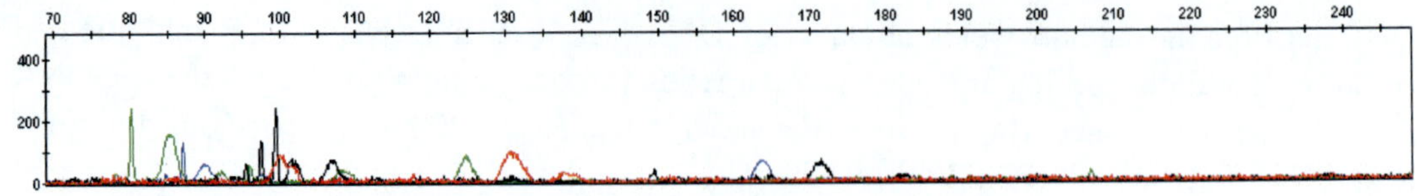

b*) PrepFiler™Forensic DNA Extraction Kit* (Applied Biosystems, USA)

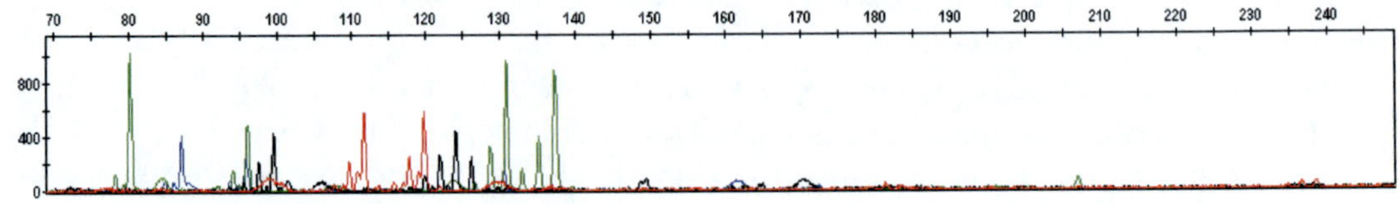

c) phenol/chloroform DNA extraction and additional extraction protocol using the *PrepFiler™Forensic DNA* Extraction Kit (Applied Biosystems, USA)

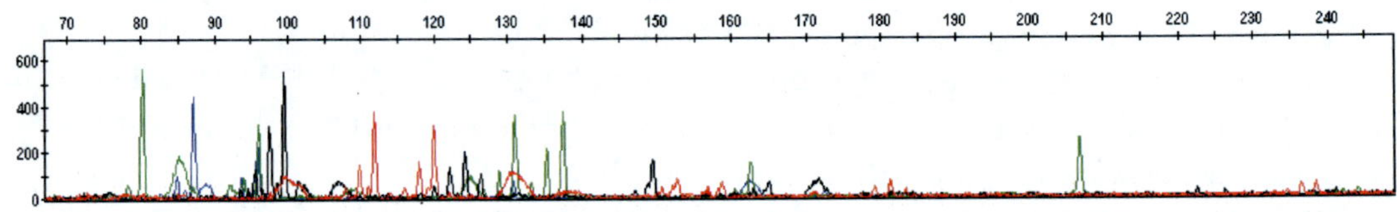

d) Same procedure as c), followed by DNA purification using *FAM PCR&GEL Purification kit* (FAMBiotech s.r.l.)

Fig.1. Bone sample DNA profile obtained with different procedures. Y-axis represents RFU (relative fluorescence units) values, X-axis represents size estimates of the fragments in base-pairs. Capillary electrophoresis separates microsatellite markers by colour and size: VLH20, HTG4, AHT4, HMS7 (blue); HTG6, AHT5, HMS6; ASB23; ASB2 (green); HTG10, HTG7, HMS3, HMS2 (yellow); ASB17, LEX03, HMS1, HMS2 (red).

Urine Samples

DNA profiling from urine is normally used in human forensics (Tsongalis et al., 1996; Marques et al., 2005), but is not well recognised in equine analysis (Marklund et al., 2006).

DNA extraction from horse urine is problematic and results are unpredictable, due to the small amount of DNA present and its rapid degradation in the sample and also because of the presence of PCR inhibitors (Khan *et al.*, 1991; Nakazono et al., 2005;Yasuda et al., 2003; Prinz et al., 1993).

Materials and Methods

DNA Isolation

Analysis was performed on a total of 50 samples which were stored for 20 days at -20°C and then refrigerated at +4°C for 2 weeks. Ten ml of horse urine, the minimum required for successful DNA profiling (Tobe et al., 2006), was centrifuged at 3000xg for 5 min and the resulting pellet was washed with phosphate-buffered saline (PBS). DNA was extracted using a PrepFiler™Forensic DNA Extraction Kit (Applied Biosystems, USA), in accordance with the manufacturer's protocol.

The DNA yield was evaluated by spectrophotometric analysis at 260 nm, using a NanoDrop ND-1000 Spectrophotometer (NanoDrop Technologies, Inc., USA). DNA concentration ranged from 0.2 to 20 ng/µl. The quality of DNA was evaluated by estimating the ratio of absorbance at 260 nm and 280 nm: 60% of the tested samples have an average value in the range 1.68 to 1.88.

To avoid potential contamination, extraction blanks containing everything except urine sample were carried through the entire process. The extraction blanks were negative for STRs amplification.

Microsatellites were amplified with Equine Genotypes™ Panel 1.1 kit (Finnzymes Diagnostics, Finland), in accordance with the manufacturer's protocol. PCR was performed in GeneAmp PCR System 9700 (Applied Biosystem, USA).

PCR products were separated by capillary electrophoresis on ABI PRISM 3130*xl* Genetic Analyzer (Applied Biosystems, USA). Size analysis of DNA fragments was performed with GeneMapper Software v3.0 (Applied Biosystems, USA).

The internal standard Genescan-LIZ 500 was used for sizing alleles. In addition, samples of known genotypes from ISAG comparison tests were used as reference to standardize the alleles.

Experiments were performed according to the ISAG guidelines (Budowle et al., 2005).

Results And Discussion

Good results have been obtained without using a purification protocol after the DNA extraction. We consider a key factor the storage conditions: in fact, the urine was stored for one month at -20°C in a sealed bottle.

The number of markers was greater than 8 in 60% of the samples (as shown in **Table 1**).

Table 1

Total number of samples = 50	
Samples without amplification	**20 (40%)**
Samples with a number of markers ranging from 8 to 11	**12 (24%)**
Samples with a number of markers ranging from 12 to 17	**18 (36%)**

No markers were obtained in 40% of samples. This was due to the presence of a high volume of acidic fluid containing a large quantity of salts and other metabolites that interfered with the PCR amplification. In addition, different amounts of epithelial cells in urine could also contribute to the negative results.

A minimum of 8 markers are needed for genetic identification, with the probability being greater than 1 in 10^8 of any two unrelated horses having identical DNA profiles (Bowling, 2001). For parentage verification extra markers are required for a statistically significant outcome.

DNA typing from urine samples could be useful after anti-doping testing, to remove any doubts of sample substitution and give unequivocal identification of the horse that provided the sample (Hawthorne et al., 2009).

REFERENCES

Azor, P., Valera, M., Gómez, M. & Goyache, F. et al. (2007). Genetic characterization of the Spanish Trotter horse breed using microsatellite markers. *Genetics and Molecular Biology., 30*, 37-42.

Binns, M. M., Holmes, N. G., Holliman, A. & Scott, A. M. (1995). The identification of polymorphic microsatellite loci in the horse and their use in thoroughbred parentage testing. *British Veterinary Journal, 151(1)*, 9-15.

Bowling, A. T. (2001). Historical development and application of molecular genetic tests for horse identification and parentage control. *Livestock Production Science, 12*, 111-116.

Bowling, A. T., Eggleston-Stott, M. L., Byrns, G., Clark, R. S., Dileanis, S. & Wictum, E. (1997). Validation of microsatellite markers for routine horse parentage testing. *Animal Genetics, 28*, 247-252.

Brevnov, M. G., et al. (2009). Validation of the PrepFilerTM forensic DNA extraction kit for extraction of genomic DNA from biological samples, *Journal of Forensic Science., 54*, 599-607.

Bruzzone, A., Iamartino, D., Blasi, M. & Pilla, F. (2003). The Pentro Horse: genetic characterization by microsatellites markers. *Italian Journal of Animal Science, 2*, 223-203.

Budowle, B., Garofano, P., Hellman, A., Ketchum, M., Sree, S., Kanthaswamy, S., Parson, W., van Haeringen, W., Fain, S. & Broad, T. (2005). Recommendations for animal DNA. *International Journal of Legal Medicine, 119(5)*, 295-302.

Davoren, J., Vanek, D., Konjhodzic, R., Crews, J., Huffine, E. & Parsons, T. J. (2007). Highly effective DNA extraction method for nuclear STR testing of skeletal remains from mass graves. *Croatian Medical Journal, 48*, 478-85.

Dimsoski, P. (2003). Development of a 17-plex Microsatellite Polymerase Chain Reaction Kit for Genotyping Horses. *Croatian Medical Journal, 44(3)*, 332-335.

Ellegren, H., Johansson, M., Sandberg, K. & Andersson, L. (1992) Cloning of highly polymorphic microsatellites in the horse. *Animal Genetics, 23*, 133-142.

Garofano, P., Veneroni, R., Molinari, S., Bonuglia, M., De Iuliis, P. & Montana, M. (2004). High throughput DNA isolation and STR analysis for animal identification: forensic casework and DNA database applications in animal genetics. *Proceedings of the 15th International Symposium on Human Identification*-Phoenix, Oct 4-7.

Gaytmenn, R., Sweet, M. Sc. & D. (2003). Quantification of forensic DNA from various regions of human teeth. *Journal of Forensic Science., 48(3)*, 622-5.

Giacomoni, E. H., Ferna´ndez-Stolz, G. P. & Freita, T. R. (2008). Genetic diversity in the Pantaneiro horse breed assessed using microsatellite DNA markers. *Genetics and Molecular Research, 18*, 261-70.

Gill, P., Kimpton, C. P., Urquhart, A., Oldroyd, N., Millican, E. S., Watson, S. K. & Downes T. J. (1995). Automated short tandem repeat (STR) analysis in forensic casework - a strategy for the future. *Electrophoresis, 16*, 9, 1543-1552.

Gjertson, D. W., Brenner, C. H., Baur, M. P., Carracedo, A., Guidet, F., Luque, J. A., Lessig, R., Mayr, W. R., Pascali, V. L., Prinz, M., Schneider, M. P. & Morling, N. (2007). ISFG: Recommendations on biostatistics in paternity testing. *Forensic Science International: Genetics*, *1*, 223-231.

Goldstein, D. B. & Schlotterer, C. (1999). *Microsatellites: Evolution and Applications,* Oxford University Press, Oxford, UK.

Guérin, G., Bertaud, M. & Amigues, Y. (1994). Characterization of seven new horse microsatellites: HMS1, HMS2, HMS3, HMS5, HMS6, HMS7 and HMS8. *Animal Genetics, 25*, 62.

Hawthorne, P., Wang-Holmes, J., Cawdell-Smith, J. & Trezise, A. E. O. (2009). DNA Profiling of horse urine samples to confirm donor identity. *Rural Industries Research and Development Corporation.*

Hoffmann, I., Marsan, P. A., Barker, J. S. F., Cothran, E. G., Hanotte, O., Lenstar, J. A., Milan, D., Weigend, S. & Simianer, H. (2004). New MoDAD marker sets to be used in diversity studies for the major farm animal species: recommendations of a joint ISAG/FAO working group. In: *Proceedings of 29th International Conference on Animal Genetics*, Tokyo, Japan.

Junge, A., Steevens, M. & Madea, B. (2002). Successful DNA typing of a urine sample in a doping control case using human mitochondrial DNA analysis. *Journal of Forensic Science., 47(5)*, 1022-1024.

Kalmár, T., Bachrati, C. Z., Marcsik, A. & Raskó, I. (2000). A simple and efficient method for PCR amplifiable DNA extraction from ancient bones. *Nucleic Acids Research., 28(12)*, e67.

Khan, G., Kangro, H. O., Coates, P. J. & Heath, R. B. (1991). Inhibitory effects of urine on the polymerase chair reaction for cytomegalovirus DNA.*Journal of Clinical Pathology*. 44, 360-365.

Lear, T. L., Brandon, R. & Bell, K. (1999). Physical mapping of ten equine dinucleotide repeat microsatellites. *Animal Genetics, 30*, 3, 235.

Loreille, O., Diegoli, T. M., Irwin, J. A., Coble, M. D. & Parsons, T. J. (2006). High efficiency DNA extraction from bone by total demineralization. *Forensic Science International: Genetics*; *1*, 191-5.

Marklund, S., Ellegren, H., Eriksson, S., Sandberg, K. & Andersson, L. (1994). Parentage testing and linkage analysis in the horse using a set of highly polymorphic microsatellites. *Animal Genetics, 25*, 19-23.

Marklund, S., Sandberg, K. & Andersson, L. (1996). Forensic tracing of horse identities using urine samples and DNA markers. *Animal Biotechnology., 7(2)*, 145-153.

Marques, M. A. S., Damasceno, L., Pereira, H., Caldeira, C., Dias, B., Vargens, D., Amoedo, N., Volkweis,R.,Viana, R., Rumjanek, F. & Neto, F. (2005) DNA typing: An accessory evidence in doping control. *Journal of Forensic Science. 50(3)*, 587-92.

Milos, A., Selmanovic, A., Smajlovic, L., Huel, R. L., Katzmarzyk, C. & Rizvic, A. et al.(2007). Success rates of nuclear STR typing from different skeletal elements. *Croatian Medical Journal., 48*, 486-93.

Nakazono, T., Kashimura, S., Hayashiba, Y., Hara, K. & Miyoshi, A. (2005). Successful DNA typing of urine stains using a DNA purification kit following dialfiltration. *Journal of Forensic Science., 50(4)*, 860-864.

Prinz, M., Grellner, W. & Schmitt, C. (1993). DNA typing of urine samples following several years of storage *International Journal of Legal Medicine., 106(2)*, 75-9.

Rohland, N.& Hofreiter, M. (2007). Comparison and optimization of ancient DNA extraction *BioTechniques, 42*, 343-352.

Tobe, S., Reid, S. & Linacre, A. (2006). Successful DNA typing of a drug positive urine sample from a race horse . *Forensic Science International., 173(1)*, 85-86.

Tozaki, T., Kakoi, H., Mashima, S., Hirota, K. I., Hasegawa, T., Ishida, N., Miura, N., Choi-Miura, N. H. & Tomita, M. (2001). Population study and validation of paternity testing for Thoroughbred horses by 15 microsatellite loci. *Journal of Veterinary Medical Science., 63*,1191-1197.

Tsongalis, G. J., Anamani, D. E. & Wu, A. H. B. (1996). Identification of urine specimen donors by the PM+DQA1 amplification and typing kit. *Journal of Forensic Science, 41*, 1031-1034.

Usha, A. P., Simpson, S. P. & Williams, J. L. (1995). Probability of random sire exclusion using microsatellite markers for parentage verification. *Animal Genetics, 26*, 155-161.

Yasuda, T., Iida, R., Takeshita, H., Ueki, M., Nakajima, T., Kaneko, Y., Mogi, K., Tsukahara, T. & Kishi, K. (2003). A simple method of DNA extraction and STR typing from urine samples using a commercially available DNA/RNA extraction kit. *Journal of Forensic Science., 48(1)*, 108-110.

In: Horses: Biology, Domestication, and Human Interactions ISBN: 978-1-61761-303-6
Editor: James E. Leffhalm

Chapter 10

Strategic Research Plan Wild Horse and Burro Management: The Bureau of Land Management, Wild Horse and Burro Program U.S. Department of Interior*

United States Geological Survey

I. Statement of the Challenge

A. Introduction Background

The Wild Free-Roaming Horses and Burros Act of 1971 (Public Law 92-195) provided protection for all wild horses and burros on federal lands and provided guidance for their management as a wildland species. At the time the Act was passed, a roughly estimated

* This is an edited, excerpted and augmented edition of a Bureau of Land Management, Wild Horse and Burro Program U.S. Department of Interior publication.

17,000 wild horses[1] occupied federal lands designated for the protection. Since 1971, the primary responsibility for management of the wild equids on federal lands has fallen primarily to the Bureau of Land Management (BLM), with the U.S. Forest Service sharing some responsibility. In 1971, the BLM was neither staffed nor prepared to monitor and manage such a large number of horses and their effects on semiarid ecosystems. By 1980, the number of wild horses had increased to 65,000-80,000 animals.

Following passage of the Act in 1971, limits for the number of horses on each herd unit (referred to as Herd Management Areas or HMA's) were set. These limits or population goals are referred to as Appropriate Management Levels or AMLs. From 1980 and through the present, more active management reduced wild horse numbers closer to AMLs. Wild horse numbers were reduced to about 40,000 by 1999 and to about 37,186 wild horses in 2003, prior to the foaling season. This number still substantially exceeds limits recommended by the BLM. Aggressive efforts are planned to reduce numbers to the established AMLs during the next few years.

Some 203 HMAs are managed by the BLM across the Western U.S. The responsibility of monitoring range conditions; allocating range resources between horses, livestock, and wildlife; monitoring horse and burro numbers and managing their population levels represents a large federal management responsibility. The number of animals in most herds are counted or estimated every three to four years, in order to plan for any gathers, and adoptions. Herd management activities, such as gathers and removals, take place for most herd areas every four years.

Mandates

The Wild free-roaming horses and burros act of 1971

In this Act, Congress stated that wild free-roaming horses and burros were living symbols of the historic and pioneer spirit of the West; that they contributed to the diversity of life forms within the Nation and enrich the lives of the American people; and that these horses and burros were fast disappearing from the American scene. Congress mandated that wild free-roaming horses and burros be protected from capture, branding, harassment, or death. To

[1] Since no systematic surveys were conducted at this time, some authors feel this number may not be representative of the numbers in 1971.

accomplish this, these animals were to be considered, on public lands where they were found in 1971, an integral part of the natural system.

Other statutory regulations

The Taylor Grazing Act of 1934 authorized the formation of the Grazing Service (subsequently the BLM) and empowered that Service to responsibly manage grazing pressures on federal rangelands. This act accelerated the capture and removal of the wild equids, which were primarily used as pet food at that time.

The 1959 Wild Horse Annie Act (named after Mrs. Velma Johnston) prohibited hunting or harassment of wild horses on public lands using motorized vehicles or aircraft.

The Federal Land Policy and Management Act of 1976 (FLPMA, PL-43, 1701) directed the BLM to scientifically manage rangelands under the principles of use and sustained yield. Under FLPMA, wild horses and burros were one of several multiple uses (along with recreation, mining, domestic grazing, fish and wildlife) that the BLM must manage in combination to best meet the public's present and future needs. Sustainability implies the "maintenance in perpetuity" of yields from public lands.

The Public Rangeland Improvement Act of 1978 (PRIA) amended PL-92-195, and defined excess horses, mandated research and provided guidance for titles for adopted horses and the adoption process.

The Stake Holders

The Bureau of land management (BLM)

The BLM's role is to manage wild horse and burro populations on western rangelands for the enjoyment of the American public. In order not to let horse and burro numbers threaten other species, or their own welfare, BLM set animal limits based on habitat and forage, and to monitor effects of various management levels on soil, vegetation, wildlife, and other ecosystem components. Grazing on public domain was primarily unregulated until 1934, at which time the Taylor Grazing Act empowered the BLM to responsibly manage grazing levels on these lands. The BLM's Wild Horse and Burro (WH&B) Program is directed from a central policy office (Washington Office, WO) in Washington, D.C., and an operations office (National Program Office, NPO) in Reno, Nevada. The BLM's 65 full- time equivalent employees, responsible for the management of wild horses and burros are supervised by decentralized operating levels of eleven state offices and several district or field offices in each state, each of which is managed by a state director.

The U.S. Geological survey, biological resources discipline

The U.S. Geological Survey (USGS) was established by the Organic Act of 1879 to provide geological, topographic and hydrologic information to the nation. This information includes maps, databases, and reports of the analysis and interpretation of water, energy, and mineral resources, land surfaces, geologic structures, natural hazards, and dynamic processes of the earth.

In 1996, the then National Biological Service was incorporated into the USGS as their new "Biological Resources Discipline" with primary goals to (1) assess and report on the nation's biological resources; (2) characterize natural processes and identify factors that influence the nation's biological resources at all levels of biological organization; (3) facilitate sound management with agency partners; (4) provide leadership in developing a biological information infrastructure; and (5) integrate ecological research, inventory and monitoring efforts throughout the USGS.

The USGS-BRD has the capability as the primary research agency within the U.S. Department of Interior not only to gather much of the essential data and to conduct valuable, long-term research for the BLM, but also to engage specialists from other agencies and institutions and to develop interdisciplinary teams to generate the high-quality science needed for informed, science-based management decisions. In late 2000, the primary role for the coordination of USGS-BRD wild horse and burro research was delegated to the Fort Collins Science Center (USGS-BRD) located in Fort Collins, Colorado.

The USDA animal and plant health inspection service (APHIS)

Founded in 1862 by President Abraham Lincoln, the United States Department of Agriculture (USDA) mission includes working for a healthy and productive nation in harmony with the land. In this context, the mission of the APHIS includes protecting America's animal resources by monitoring and managing animal disease conditions existing in the United States, resolving trade issues related to animal health, and ensuring the humane care and treatment of animals.

An advisory relationship was created between the BLM and APHIS in February 1999 to allow APHIS to provide assistance, consultation, and coordination on issues relating to the health and proper handling of wild horses and burros that are under the management of the BLM. Through a reimbursable agreement, the APHIS receives needed budgetary support for the wild horse and burro activities of its field force who facilitate the agency's ability to respond to emergency disease outbreaks and provide animal health monitoring and surveillance.

Domestic livestock grazers and wildlife groups

Domestic cattle, sheep, and goats graze many of the federal lands where wild horses and burros are managed. Livestock grazers often feel that wild horses and burros compete with domestic livestock for range forage. Traditional sportsmen and wildlife agencies may also feel wild equids encroach upon their interests to produce harvestable big game from the same areas. These species likely do compete and conflict with each other in some situations, but the BLM's role is to minimize these conflicts. Wild horses and burros are provided habitat on public lands and it is the challenge to the BLM to obtain the common objective of balance among resource users.

Wild Horse advocates

Wild horses and burros have a considerable following in the American public who consider the animals to be part of their western heritage and aesthetic enjoyment of the federal lands. A number of advocacy groups maintain that wild horses and burros should receive the first preference in any conflicts with other resource uses on lands where they are protected.

The Strategic planning process

In late 2000, the Fort Collins Science Center of the US Geological Survey, Biological Resources Division (USGS-BRD) was charged with developing a strategic research plan for the management of wild horses and burros. Later, contributions by APHIS were incorporated into the research planning process to help address issues related to health and handling. The purpose of this strategic planning process was to:

- review past progress and identify problems;
- set broad goals central to the BLM's mission for wild horse and burro management;
- establish specific, time-bound, measurable goals, and strategies to achieve them; and evaluate the progress towards those goals at set time periods, and to readjust the planning as needed.

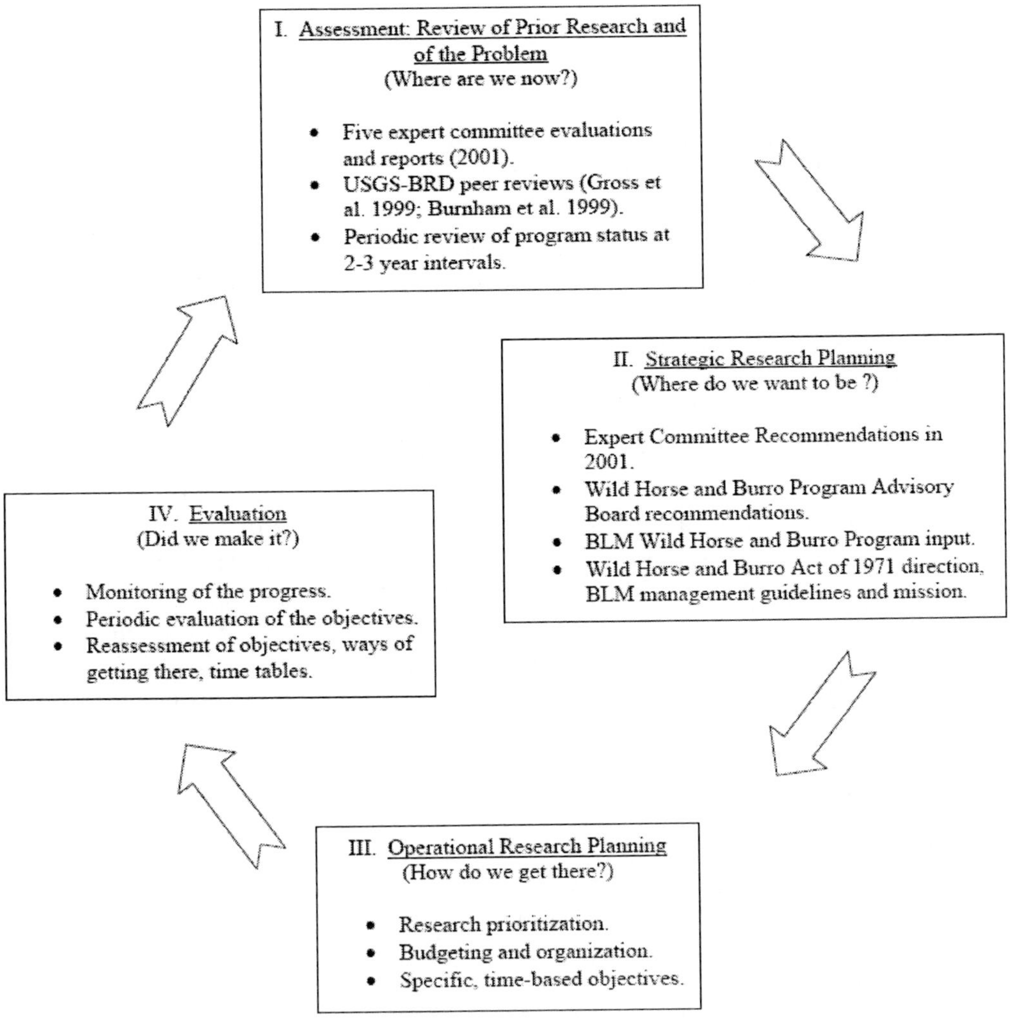

Figure 1. Proposed comprehensive strategic research planning process for wild horse and burro management for the Bureau of Land Management, 2001-2010.

Strategic plans are long-range, broad-based documents that create a specifically stated vision of the future. Strategic planning is the long-term directional planning that occurs at the highest level of the program (in this case, wild horse and burro research) and determines the overall success or failure of the program. Strategic plans are one functional segment of a Comprehensive Management System (Organization of Wildlife Planners 1997). First, an inventory and assessment is made of the current state of the research, the client's and stakeholder's needs, and the driving policies and legislative mandates (Figure 1). Second, a strategic plan is developed by the agency in collaboration with the other stakeholders. It details roles, responsibilities, objectives and the tasks that must be accomplished to arrive at the objectives. Third, an operational plan is developed to clarify and quantify the tasks, and outline the process to achieve the objectives. Last, periodic evaluation points are established to monitor the success of attaining clearly stated performance measures and performance standards. The strategic plan is intended to be modified to the extent necessary, at three to five year intervals, based upon these periodic evaluations of progress by the BLM. These steps are repeated until all of the objectives are met, projected in this instance to require one decade.

This strategic plan is a joint effort of the USGS-BRD, the BLM, and the APHIS. This plan was developed to provide the BLM with a research strategy to meet the needs for management and care of wild horses and burros. The plan was developed over a period of 9 months with the input of 39 subject area experts from 11 universities, 3 federal agencies (BLM, USGS-BRD, APHIS), and two state wildlife agencies. USGS-BRD took the lead role in planning and coordinating meetings of the expert committees and in drafting the strategic plan based on committee and agency inputs (compilers, F. Singer and M. Tobler). Assisting in this effort was the BLM's wild horse and burro research coordinator, L. Coates-Markle, and equine health experts from the APHIS, L. Hatcher, and A. Kane.

The strategic research planning incorporated: (1) recommendations from five expert committees in 2001[2], (2) recommendations of the BLM WH&B Advisory Board, (3) direction from the Wild Horse and Burro Act of 1971, (4) input from the BLM managers and specialists assigned to the Wild Horse and Burro Program, (5) periodic milestone decisions by the BLM based upon the research findings and direction, and (6) earlier review of BLM and BRD research (Smith et al. 1996; Gross et al. 1999; Burnham et al. 1999; Population Viability Forum 1999; National Research Council 1991). It was upon this foundation that the strategic plan was built. This strategic plan will be updated at five-year intervals.

There had been no previous prioritization of the BLM's wild horse and burro research and management needs, or no effort to develop a strategy for fulfilling those needs within a specified time period. This strategic planning process will fill this void.

Research administration and approvals

The BLM will determine the research priorities for each year with input from the USGS-BRD and the APHIS. A WH&B Research Advisory Committee (also referred to here as the Research Team) will be convened with equal participation by the BLM (the BLM's research coordinator), the USGS-BRD, and the APHIS to provide recommendations to the WH&B Group Manager. A three-member advisory committee is envisioned at this time: F. Singer, L. Coates- Markle, and A. Kane. This committee will meet twice per year to recommend

[2] The committee reports may be obtained by accessing www.wildhorseandburro.blm.gov

research priorities. One of these meetings will occur in April of each year to coordinate with the BLM's annual budget process. Advice and input will continue to come from the BLM's Advisory Board, the BLM WH&B Group Manager's staff and steering group, BLM Director's Science Advisory Committee, BLM's wild horse and burro specialists, and from any topic-specific advisory panels (seven have been convened to date) that may be convened by USGS-BRD. The Group Manager will meet with the Research Team at least once per year and will resolve any final issues of budgets, priorities, methods, coordination with field specialists, etc. The BLM, as the management agency, must make all final decisions and must set all final research priorities.

Research contracts, cooperative agreements, work orders, and interagency agreements will be administered through the USGS-BRD, Fort Collins Science Center (FORT), APHIS, or the BLM. A logical breakdown of topics will be used according to the appropriate expertise of these agencies on a case-by-case basis. Scientific expertise from the best experts in the scientific community will be sought in any subject area where these agencies and their cooperators can benefit. The APHIS will take the lead role in research dealing with health and handling issues. The USGS-BRD will take the lead role for research as it relates to the assigned topics of contraception, population estimation, genetics and populations modeling.

As the research program grows in funding, the amount of funding to outside experts will likely expand to all subject areas. For those projects conducted by USGS-BRD at FORT, all standard USGS requirements and approvals will be followed including:

1. A written study plan reviewed and approved by two-peer scientists and one statistician.
2. Approval of any animal handlings by the USGS-BRD Animal Care and Use Committee.
3. Center review of all manuscripts and proposals.

USGS-BRD will maintain and manage a core-fund of USGS base research dollars that is earmarked exclusively for wild horse and burro research. Additional USGS funding may come from in-house periodic calls for proposals (USGS Venture Capital Fund, USGS Ecosystems Studies, USGS Species of Federal Concern). USGS scientists submitting proposals will follow the specific proposal guidelines for those programs. The BLM and the USGS may also contribute research dollars in addition to these USGS base dollars.

The competitive call-for-proposals process will be the preferred mechanism for most contracted research, although sole source contracts or university cooperative agreements may be to the advantage of the government in some situations. The Research Team will prepare standard guidelines for the preparation of proposals over the next few months. All proposals assigned to the USGS-BRD for project management must follow USGS contracting requirements that mandate that a panel of qualified specialists rank the proposals. The review panels will include agency representation from BLM, APHIS, and USGS. In addition, more specialized subject experts may be added to these panels.

The Contracting Officer (CO) for the contracts will be the respective agency's regional Contract Officers. The CO Representatives (COR), Project Inspectors, and Technical Advisors will be drawn from the pool of subject area experts from within USGS-BRD, FORT, the APHIS, and the BLM. All these assignments will be determined on a case-by-case basis and may include BLM field resource managers. All unsolicited proposals will also need to follow the same standard guidelines. Some unsolicited proposals may be funded with outside funds, but they still need to fit into the general scope of the BLM's needs, as identified in the planning process, and they should follow the same guidelines for drafting the proposal. Animal Care and Use and other appropriate approval processes must be met exactly as in the case for solicited proposals. Occasionally, exceptional unsolicited proposals will be funded from core research funds. University contractors must obtain the approval of their own university Animal Care and Use Committees before their research proposals may receive final funding approval. For those research organizations with no such approval process, either the APHIS or the USGS-BRD Animal Use and Care Committee will review the handling and care planned, depending upon which agency has been assigned the management of that particular project.

B. Wild Horses in North America

History

The wild horses that roam the west are feral descendents of domestic stock brought to North America by European colonists. No native wild horses existed in the Americas at that time, even though the horse evolved in North America and spread to Eurasia approximately 2.53.0 million years ago. The North American fossil record suggests that progenitors of all extant horses, asses, and zebras once lived in North America.

The last remaining native wild horses persisted in North America until as recently as 8,000-10,000 years ago when they mysteriously became extinct. Recent paleontological finds from Alberta indicate that these last remaining small native wild horses were killed and eaten by Native Americans about 10,000 years ago. Perhaps over-exploitation by Native Americans in this predomestication period played a role in the horse's demise in North America. Climate change and changes in vegetation have likely also played a role (Hulbert 1993, Martin and Klein 1984, Sharp and Cerling 1998, McFadden 1992). The disappearance of the native form of such an adaptable and widespread species as the wild horse from North America several thousand years ago remains an enigma.

The progenitor of the domestic horse (*Equus caballus*) which was domesticated roughly 6,000 years ago, is suspected to have been a tarpan-like animal—a short, stocky, mousy or yellowish gray (possibly dun or grulla) animal about the size of a large pony. The tarpan persisted into the early to mid 1 800s in western Europe and the Ukraine where the last animal was shot in 1879. The tarpan also did not survive in captivity, the last one died in 1918, although the closely related Mongolian or Przewalski wild horse (*Equus-caballus przewalskii*) did survive in captivity. The Przewalski horse has a different chromosome number and thus, is not the progenitor of the domestic horse (Bennett and Hoffman 1999). Representative examples of tarpan-like animals were reconstructed by breeding the last remaining captive tarpan domestic mixed horses. Tarpan-like animals have recently been released into the wild in several reserves in western Europe. Przewalski horses have also recently been released in the wild into Mongolian reserves.

Horses were reintroduced in North America by the Spaniards during the 1500s. The Spanish or Iberian influence remains very strong in some of the wild horse populations that have the longest histories of escape from domestication (e.g., the Kiger, Pryor Mountain, and

Sulphur Mountain herds have strong Spanish ancestries). A number of U.S. western breeds were derived from these earliest Spanish bloodlines (the Spanish Mustang, the Rocky Mountain horse, the Choctaw horse; Bennett 1998). Later however, military, saddle, and draft, stock horses dominated by the Thoroughbred, Morgan, Quarter horse and draft breeds escaped into the western rangelands or were intentionally released and rapidly increased, forming broad zones of introgression with the earlier Iberian Spanish colonial bloodlines. These wild horses of mixed ancestry eventually increased to very large populations that inhabited vast areas of the U.S. western rangelands. Only a small number of the wild horse herds retain a largely Spanish colonial ancestry.

The Management challenge

Survival rates for wild horses on western public lands are high. None of the significant natural predators from native ranges of the wild horse in Europe and Asia — wolves, brown bears, and possibly one or more of the larger cat species — exist on the wild horse ranges in the western United States (mountain lions and black bears take foals in a few herds, but predation contributes to population limitation in only a handful of herds (e.g., Montgomery Pass). In some cases, adult annual survival rates exceed 95% and many horse herds grow at sustained high rates of 15-22% per year. Prior to herd management, many wild horse ranges were overgrazed, and some wild horses died from malnutrition or dehydration.

Although wild horses occur in 10 states, the vast majority of the animals are located in Nevada (74%) and Wyoming (10%). The BLM has established appropriate management level (AML), or population goals, based on range conditions and monitoring data, while the AML setting process is still underway in other management areas. Current numbers of wild horses are substantially in excess of these goals.

A major challenge for the BLM wild horse managers is what to do with the large number of excess horses produced each year on western rangelands (see Wagner 1983 and National Research Council 1991 for excellent reviews). Once instituted, management in most herds typically involves helicopter gathers and adoption of excess animals. Most horses and burros are adopted by the public, but not all of them can be placed in homes. Older adults, such as some of the mature stallions, are not attractive as potential animals to adopt. Many unadoptable animals are kept for years in captivity in long-term pasture holding facilities. Research into the use of contraceptives to limit the growth of wild horse herds has been ongoing since the 1 970s, both in herds on western rangelands and on several eastern barrier islands. Four of these herds on eastern barrier islands are currently managed with immunocontraceptive agents. Tests with immunocontraceptives have been conducted on a few of the larger wild horse herds in Nevada. However, no free ranging western horse herds have yet been managed at the population level with contraceptives.

Several strong and diverse public interest groups pressure the U.S. Department of the Interior regarding the management of wild horses on public lands. These interests include, at one extreme, those who strongly support the protection and management of wild horses with little or no human intervention. At the other end of the spectrum are those who favor intense management of wild horses and burros, with an objective to maintain very low numbers of horses and burros. Elements of the public, such as those that pursue domestic livestock grazing and the harvest of big game wildlife, may view wild horses and burros as some competitors for other resource uses.

The BLM requires the highest quality science to simultaneously manage for sustainable production of all components of the ecosystem and for healthy, free-roaming, and genetically viable populations of wild horses and burros.

History of contraceptive research

The Problem

The BLM's need for a fertility control agent to manage the numbers of wild horses was recognized soon after passage of the Act in 1971. The federal government sought to reduce the number of animals that they needed to gather, transport, adopt, feed and maintain (in the case of non-adoptable animals).

Testosterone propionate in stallions

In 1978, the BLM entered into a series of research contracts that continued into the 1 980s focusing primarily on development of a chemosterilant for wild stallions. The primary hormonal agent tested, testosterone propionate, was found to be effective in reducing sperm motility and effectively sterilizing the wild stallions for up to six months (Garrott and Siniff 1992). The hormone did not affect libido of the stallions, but foal production was reduced 83% in harems with treated stallions compared to those with untreated stallions (Kirkpatrick and Turner 1982).

There were a number of drawbacks to use of the hormonal sterilant. Application required immobilization of the stallion and injection of a large dose of the agent. A dose consisted of 2.5 - 10.0 g of a long-acting form of testosterone propionate, encapsulated in a biodegradable, nontoxic lactide coating to prolong the release of the hormone over a four to six month period. Additionally, (1) foaling times in treated herds shifted into the summer or fall (Garrott and Siniff 1992); (2) herding, capturing, immobilizing and injecting the harem stallions on an annual basis was difficult and costly; (3) long-term effects on the treated horses remained unknown; and (4) it was believed the agents could enter the food chain. For these reasons, chemical sterilization of stallions was abandoned as a focus for research into fertility control.

Silicone Implants in mares

Silicone rods impregnated with progesterone and estradiol were implanted into the necks of mares and blocked ovulation for up to 28 months (National Research Council 1991). This work was also suspended because the invasive nature of the surgery and the unacceptable stress placed on mares.

The PZP investigations

The scientific community identified the needs for an ideal fertility control agent in 1991 as follows (Kirkpatrick and Turner 1991; Seal 1991):

1. The agent should be at least 90% effective.
2. The agent should be capable of administration by remote delivery.
3. The agent should either be immediately reversible, or its effects should passively wear off.
4. The agent should be safe to pregnant animals.
5. The agent should not pass through the natural food chain.

6. The agent should be inexpensive.
7. There should be no debilitating side effects on the health of the horses.
8. The agent should not influence the social behavior of the horses.

This list of needs would drive much of the U.S. contraceptive research into wildlife species during the 1 990s, including research funded by both the BLM and the USGS-BRD. To meet the stated criteria, the National Park Service (NPS) research team on Assateague Island National Seashore turned to an immunocontraceptive agent, porcine zona pellucida (PZP), for the wild horses[3] on the island (Kirkpatrick and Turner 1982; Kirkland et al. 1992; Turner 2000; Kirkpatrick 1995), which had been reported to block fertilization in dogs, rabbits, and primates. In order for sperm to attach to the ovum and fertilize the egg, there must be complementary proteins on both the surface of the sperm and the zona pellucida (ZP) of the ovum. PZP is a foreign protein against which the treated mare produces anti-PZP antibodies. These antibodies attach to the mare's zonae sperm receptors on the ovum and block fertilization (Floorman and Wasserman 1985; Kirkpatrick 1995). Zona pellucida from domestic pig ovaries (obtained from slaughter houses) is minced and the PZP is obtained from screening filtration. Freund's Complete Adjuvant (FCA) is mixed with the PZP in order to enhance its effects when it is initially injected into mares intramuscularly.

Experimental PZP application on the wild horses of Assateague Island began in 1988. Following promising reductions in the pregnancy rates in mares (Kirkpatrick 1985; Kirkpatrick et al. 1990), the NPS in 1994 began to stabilize the growth of the population solely using PZP immunocontraception (Kirkpatrick and Turner, in press). The Assateague research team also developed non-invasive methods to assess the pregnancy rates of, and detect ovulation in, free- ranging treated and non-treated mares by analyzing reproductive steroid metabolites in feces and urine (Kirkpatrick et al. 1992). These methods require the sample be taken in the field from individually recognizable mares, but no captures are necessary.

It is impossible in this report to do justice to the larger volume of research conducted on PZP in wild horses on both western rangelands and on the eastern national seashores. Instead, we direct the reader to the summary papers of the J. Turner/J. Kirkpatrick/I. Liu research team (Kirkpatrick et al. 1992; Kirkpatrick 1995; Turner et al. 2000; Kirkpatrick and Turner, in press; Turner and Kirkpatrick, in press), to review the papers of Kreeger (1997) and Curtis and Warren (1999) and to annual reports of the research study (Turner et al. 1993 through 2001). Copies of these reports may all be obtained by writing to the National Program Office, Wild Horse and Burro Program, Reno Nevada.

The Outlook for PZP

The PZP agent appears to meet most of the safety concerns of the BLM: it does not enter the food chain, its effects passively wear off with time if the injections are terminated, normal reproduction can be resumed, following up to seven years of use, and it does no harm if injected into mares that are already pregnant — they carry foals to term. Initial research suggests native PZP does not affect ovarian function, hormonal health, or safety in pregnant animals (Turner et al. 1999, 2000; J. Turner, personal communication, Nov. 26, 2001). Life

[3] These animals are small horses, not ponies. Horses and ponies differ in several important ways, such as in length of gestation.

span and health of treated mares may be increased, apparently due to the absence of stresses from pregnancy and lactation. Treated mares apparently live about five to ten years longer than do untreated mares that continue to get pregnant and produce young (Kirkpatrick and Turner 2002). One initial study suggested harem behaviors are not influenced (Powell 1999). There appear to be no generational effects — offspring of treated mares are able to reproduce normally (Kirkpatrick and Turner, in press; Turner and Kirkpatrick, in press). The agent is about 90% effective in blocking fertility in mares.

Best results using PZP are achieved following an initial "primer" dose, followed by annual "booster" shots. The initial injection, or primers, may be administered to mares following gathers when they are in chutes during capture. Alternatively, in those populations where the individual mare can be both recognized and approached on foot for darting, the injection may also be administered remotely by means of a 1.0cc dart with a Pneu-bait or Dan-Inject dart gun. A second booster shot is then required for each year of immunocontraception. Following the second or third year of treatments, only an every-other or every-third year booster is needed (J. Kirkpatrick, pers. comm.). Following cessation of the annual treatments, the agent and the antibodies passively decline, anti-fertility effects wear off, and normal reproductive function is resumed the subsequent year. However, following seven or more years of treatment, the antifertility effects may be permanent for individual mares (Kirkpatrick and Turner 2002).

PZP has been successfully applied to control fertility and limit the size of several small populations of wild horses on eastern barrier islands for periods of a few to 14 years (Assateague Island National Seashore, Cape Lookout National Seashore, Shackleford Banks; Carrot Island, Rachel Carson National Estuarine Reserve; and Little Cumberland Island, a private island). Progress is continuing on development of a time-release pellet vaccine of PZP that will allow two years (actually ~22 months) of fertility control with only a single shot injection (Turner et al. 1999, 2000, 2001). Progress on this time-release form is encouraging, although efficacy rates are variable and may be slightly lower (~ 85%) than for the conventional multiple injection program.

Two major drawbacks of conventional PZP and Time-Release PZP have been identified by BLM managers: (a) the brief duration – managers prefer a one-shot, three to five-year duration, and (b) the fact that the most effective known adjuvant, FCA, present some health concerns. While the one-year or two-year durations of these forms may be adequate, and even preferred for small populations of wild horses, managers of the larger herds, such as herds in Nevada and Wyoming, have a critical need for a single application agent that lasts longer. If a gather is held during the summer or early fall, and the Time-Release PZP is injected, only one effective season of contraception maybe achieved. Also, some mares could become pregnant late during the second subsequent summer. There are some concerns about a lower survival of late born foals.

The BLM's long-term management needs clearly include both a shorter duration agent for small herds and a longer duration agent for the largest herds. A high priority of this research plan is to develop either a further time release extension of PZP, or develop a new agent that meets this need for a single-dose, longer duration contraceptive. In response to these needs BLM has just implemented a long-term study with captive wild horse mares to test the safety and efficacy of a newly-developed 3-4 year PZP vaccine.

BLM also seeks an alternative adjuvant to FCA. FCA can cause health problems when used in horses and following accidental injection or needle stick in people. FCA causes a false positive TB test, and can cause granulomas at injection sites in treated mares. These are generally small and shrink over time when the injection is into the buttock area of the horse. Presently, these risks are mitigated by only allowing persons trained and certified to administer the PZP and FCA mixture. These individuals must carefully following administration protocols in the field. Zoo Montana has trained BLM specialists in the past to handle the PZP and FCA mixture. However, a safer alternative adjuvant is desired. Modified Freund's Adjuvant (MFA) and other adjuvants may be potential replacements. Preliminary results from a recent study comparing the efficacy and duration of MFA as a replacement are encouraging and suggest that MFA may be an effective replacement for FCA with wild horses. BLM will substitute a new adjuvant as soon as an effective replacement for FCA can be identified.

One intriguing new candidate agent form of PZP is referred to as SpayVac™. It is simply PZP, combined with a FDA-approved adjuvant, incorporated within or between several layers of multi-lamellar liposomes. The liposome technology was developed by a team of scientists headed up by Dr. Robert Brown of Dalhousie University, Nova Scotia. The patent is held by Dr. Brown and Mark Fraker of Terramar Environmental Research Ltd., Vancouver, B.C. The liposome technology effects are not completely understood, but in one study using harbor seals, titers to the agents remained high following a single dose and pregnancy was blocked for more than eight years in the female seals (Brown et al. 1997). One big advantage of SpayVac™ is that its main component is PZP, and thus, many PZP research results should also apply. Although the duration of SpayVac™ is thought to be dose dependent (Lowell Miller, APHIS, May 2003, pers.corres), how long the effect will last, whether the duration will qualify SpayVac™ for use by the BLM, and whether the duration of contraceptive effects can be modified by altering the current liposome mixtures are questions that remain to be answered.

II. Overview of the Issues

A. Health and Handling Issues

The BLM gathers and holds in facilities wild horses and burros for periods of weeks to several months prior to adoption. In some cases if animals are determined to be unadoptable, they may be held for several years in long-term pasture holding facilities. During these pre-adoption and long term holding periods, the animals are under the care and supervision of the BLM. Health care includes multiple inoculations against pathogens, hoof trimming as needed, and de-worming on a regular basis. The animals are wild, handling creates some stress and they may be injured during handling, transport or treatment activities. Animals are also concentrated in the facilities, and infectious diseases can rapidly pass through the animals.

The BLM seeks to minimize the stress of handling and use the most effective and cost efficient health management practices available to safeguard the health and well-being of the animals under their care. Research is needed to identify optimal handling and healthcare practices that are not completely understood at this time because of the distinct differences between wild and domestic equids and the unique challenges associated with the large facilities in which they are managed.

Health problems among free roaming wild horses and burros are uncommon. Most of the health research needs of the free roaming wild horse and burro populations under BLM management are limited to better understanding how to manage emergency situations related to severe range conditions and how range conditions influence the health of animals at the time they are gathered. The need to support research to improve our understanding of the health of wild horses and burros on- and off-the-range is recognized in the context of the overall strategic plan for wild horse and burro research.

B. Fertility Control in Wild Horses[4]

Fertility control cannot be used to reduce herds of wild horses that are substantially over AML, or alone to limit population growth. Fertility control will assist the gather and removal program in achieving these two goals. The BLM seeks two fertility control agents, or two forms of the same agent that will reduce the frequency and/or size of expensive gathers and removals, and that will reduce the number of animals that will need to be removed from the range and adopted. The BLM seeks a shorter-term duration agent that will provide two years of contraception following a single injection for safe management of smaller herds of wild horses where population viability is a concern. The current form of Time-Release PZP may meet the need of the short-term agent, pending the results of the planned field trials. For larger herds of wild horses, the BLM seeks a contraceptive agent that lasts three to five years following a single injection. Gathers occur on a four-year cycle. Contraceptive management could focus on young to prime-aged mares following genetic guidance (Gross 2000). These mares could be injected once with an agent with a four-year duration, thus contracepted for four years, but then be allowed to passively return to fertility and contribute genetically for the remainder of their lives. A new group of young or prime-aged mares would be treated at the next four-year gather.

FCA is the most tested, most efficacious, and the best available adjuvant at this time. However, recent studies suggest the agencies should expand the search for a new adjuvant.

[4] The listing order of topics does not imply priorities of timing or funding. The BLM will establish timing and funding priorities on an annual basis following input from USGS-BRD and APHIS.

There are concerns that FCA could cause some health problems. The BLM is working to replace FCA with another adjuvant, such as MFA, as soon as possible.

Sufficient prior work with PZP has been conducted on wild horses in Nevada and on Assateague Island to justify field trails at this time. However, significant unanswered questions remain concerning population and behavioral effects of the treatments that must be addressed before BLM proceeds with broad-scale management applications of fertility control.

Development of a new agent, or further time-release development of PZP, may follow one of two different paths. First, the BLM may pursue the path of Food and Drug Administration (FDA) approval of the agent. Full commercial development of the agent may be necessary, to defray costs of the approval process. A commercial company could produce the agent for purchase by the BLM within their wild horse and burro program. The amount of agent needed annually by the BLM is too small alone to elicit much commercial interest. For example, it is estimated one person in a few weeks time can produce all of the PZP that would be needed by the BLM to control wild horses each year. If the agent was also useful for other purposes (e.g., spaying of dogs, cats, or livestock) perhaps commercial interest might occur. This process can be time consuming and expensive (e.g., several million dollars), FDA approval of all safety, health, and efficacy work (Good Laboratory Practice and Good Clinical Practice guidelines must be followed) must occur, and the FDA must approve the final product. Approval of agents that already have prior usage and prior approval in domestic food animals, however, might be within the realm of possibility for the BLM. Second, the BLM may continue to use PZP or other new products under a research protocol. The BLM has selected this latter option for now for reasons of practicality and costs, although commercial development still remains an option at any time down the road. Third, the BLM may seek a legislative, non-investigational exemption for use of the current PZP mixture.

Reproduction in wild horses may differ in a number of subtle ways from domestic horses, for example, in the season of estrus cycles. The explanation for these observed differences may be merely environmental, or there may be some minor physiological differences that are inherited. Until the differences can be defined, the BLM has determined that non-invasive research into fertility control could be conducted on captured wild horses held in BLM facilities or with free roaming wild horses under the Fertility Control Field Trial Plan.

Fertility control is not as immediate and pressing a need in wild burros, as it so clearly is with wild horses. The social structure of burros, which lacks stable harem breeding units, combined with year-round breeding; would prove challenging for application of the current PZP technology. However, this challenge should not prevent the BLM from pursuing a solution to fertility control in wild burros. Given the limited financial resources available for fertility control research, it should be pursued in wild burros just as soon as some successes can be achieved with wild horses.

C. Population Estimation and Modeling

Wild horse and burro populations increase at a high annual growth rate and these high growth rates can be sustained annually for many years. Active management programs are required for wild horses and burros that include detailed tracking of population sizes, population growth rates, sex and age composition, and modeling of options for removal strategies and population goals.

Accurate population estimates for planning and management activities are essential. The management applications of either removal or contraception (or a combination of both) are based on the number of animals to be reduced and the intervals between management; however, these goals are only as accurate as the population estimate. Wild horse and burro managers need accurate and defensible aerial surveys.

A user-friendly computer model and manual has been developed to simulate the growth rate and long-term planning of management removals for wild horses (Jenkins 1996). Managers are in need of a similar model to simulate wild burro population dynamics and to plan removals. During aerial surveys made in the flat, treeless terrain of Nevada and Wyoming ~ 85% of the wild horses present are seen (Garrott et al. 1991). For aerial surveys over forested or tall shrub desert areas, there are no methods to determine the number of animals missed. One exception to this is the work of the Arizona Interagency Wild Burro Working Group, consisting of the BLM, State of Arizona, and other federal biologists. This group has developed aerial estimation techniques for wild burros in central Arizona with promising results.

D. Genetics

Although the current total number of wild horses in all the herds is very large (37,135 animals), the management goal for most herds is small. The BLM's stated AML goals are to manage 41% of the wild horse herds at a census number (N) of < 50 horses, and 54% of the herds at $N < 100$ horses. Genetic effective population size[5] for some of these herds may be set too low. At first glance, these statistics appear to be cause for concern.

Information is not available on how many of these small herds are truly isolated. If there was even occasional gene flow between two or more herds that resulted in at least one or two successful breeding animals every generation[6] that produced breeding offspring, then the genetic resources of all the groups would be maintained. Such groups of two or more subpopulations whose population dynamics are independent, but are connected by low levels of movements and gene flow, are referred to as a metapopulation. Clearly, many of the smaller wild horse populations are probably part of a larger metapopulation.

Inbreeding is apparently rare in wild horse populations. Horses in only a very small number (approximately 5) of the 203 HMAs have exhibited characteristics possibly attributable to inbreeding, such as cataract blindness, dwarfism, parrot-mouth, or club foot

---

[5] Genetic effective population size (N_e) is a measure of the number of animals within a population that are not only breeding, but their progeny are successfully contributing their genes to the next generation. This complex topic is covered in more detail in Appendix II and III. N_e can be used to estimate the rate of loss of genetic heterozygosity (H_o) and loss of allelic diversity per generation in a population. The recommendation for $N_e > 50$ came from the breeders of domestic animals that found the level of loss acceptable. This number will result in a predicted loss of 1% of the heterozygosity (H) present per generation. The $N_e = 50$ rule may not be adequate for wild populations since selection pressures are more severe in the wild. N_e of larger than 50 may be required for wild species to adapt to more severe environmental conditions or changes. The N_e calculations assume random mating, which is never true with equids, and no mating of close relatives, which might not be the case in the smallest equid populations.

[6] A wild horse generation equals 5-14 years.

deformities. Most wild horse herds that have been sampled, exhibit moderate levels of genetic heterozygosity (both allozyme or biochemical and DNA heterozygosity) (Bowling and Touchberry 1990). Thus, there does not appear to be any immediate cause for concern about inbreeding depression[7] in wild horse herds.

Nonetheless, the committee recommended a cautious approach by the BLM. The agency should monitor for signs of inbreeding via conducting genetic surveys; monitoring genetic information and key indicators of genetic health; and estimating N_e, in a representative sample of herds (about five should be adequate). High inbreeding coefficients, and any bottlenecks[8], or losses of genetic heterozygosity are a cause for concern. However, research is still needed to determine optimal goals for genetic heterozygosity that will maintain the fitness of horses in the wild herds and prevent inbreeding depression.

E. Habitat Assessment and Setting Population Goals

Central to the BLM's management of wild horses and burros is the habitat evaluation process that the BLM specialists use to set the appropriate stocking numbers of wild horses, burros, domestic livestock, and any native grazers, such as elk or bighorn sheep. For wild horses and burros, this numeric goal is the AML. Equally important is the monitoring program that the BLM uses to determine the success of these various AMLs in protecting soil, vegetation, and wildlife resources of the area.

In 1988, the Department of the Interior's Board of Land Appeals decided that the wild horse and burro stocking levels and livestock numbers be set to achieve a "thriving natural ecological balance" for each herd management area. As noted earlier, the Federal Land Policy and Management Act of 1976, the Public Rangelands Improvement Act of 1978, and orders from Congress have directed the BLM to manage the number of wild equids to accommodate multiple uses of other resources and the long-term sustainability of the range.

Diverse methods have been used to set the AMLs on the BLM lands. Currently all offices are required to adhere to the Land Use Plan process to determine AMLs and livestock numbers, including public scooping, plan development, National Environmental Policy Act analysis, and information from on-going monitoring. But since state offices have the lead role in habitat assessments, there is some diversity in how the various states conduct habitat monitoring. The BLM directs specialists to use new monitoring data to update forage allocation decisions in an adaptive management approach.

The BLM has an ongoing need for the incorporation of most up-to-date research findings and new techniques into their habitat monitoring and habitat restoration programs. For example, a GIS-based habitat model for wild horses would be useful to many managers.

The sheer size, diversity, and complexity of the BLM's current program, however, mandates that an in-depth assessment of habitat research needs to be conducted by the BLM. Such a major assessment was beyond the scope of the 2001 effort. Additionally, habitat research needs were viewed by the BLM as less pressing than other research needs (at least in 20012002). Several changes and improvements are currently underway in how the BLM deals

[7] Inbreeding depression is defined as the loss of fecundity or viability due to inbreeding. It can frequently manifest itself as a specific genetic defect in horses.

[8] A population bottleneck is defined as a short-term population reduction to a small size that may reduce genetic heterozygosity.

with all of their grazing management, including how livestock, wild horses', and burro's grazing is integrated. It was premature to propose any additional changes at this time. Habitat research is deferred until such an in-depth analysis can be conducted by the BLM.

III. The Strategic Plan

A. Health and Handling Strategies

The Challenge

While they are generally considered hardy, wild horses and burros face new challenges following capture. Even under the best conditions, simply handling wild animals to move them from pen to pen, load them on trailers or administer preventive or therapeutic medications is stressful and includes some increased risk of injury. Physiologically, stress increases susceptibility to infectious disease. Captured animals may be naive to diseases that are common among domestic equids (e.g., strangles), and they are mixed in larger groups than are typical under rangeland conditions. The logistics of gather, preparation and adoption often include the repeated transportation of animals over long distances. These factors combine to create unique health and handling challenges for the wild horse or burro compared with their domestic cousins.

Safeguarding the health and welfare of wild horses and burros includes identifying the most effective, least stressful way of handling and preparing them for their transition into captivity. A clean bill of health is also an important part of successful adoptions.

Most of the health research needs for wild horses and burros on-the-range are related to better understanding the health problems that occur in free ranging wild horses and burros. These are usually the direct or indirect result of man's influence (i.e., domestic horses exposing wild horses to infectious disease) or unusually harsh range conditions (e.g., drought, fires, plant toxicities).

Issues

1. Animals are placed in stressful circumstances during handling and transport from one holding facility to another, and illness and injuries may occur.
2. Infectious upper respiratory disease, including strangles and streptococcus zooepidemicus infection, occurs in animals in some facilities; and the sources of these infections are not known.
3. Record keeping concerning mortalities, nonfatal illness or injury, and medical and preventative treatments used at different facilities needs improvement.
4. Wild burros are unique and may have unique health problems (e.g., susceptibility to hyperlipemia and hyperinsulinema) and can become ill if moved too rapidly to areas with very different climates.
5. Optimal vaccination strategies and protocols to prevent certain infectious diseases (e.g., strangles, equine influenza, and herpes) and the effects of handling stress are not well understood.
6. We need a better understanding of how to best intervene when animals are acutely or chronically exposed to extreme range conditions brought on by drought or fire including water deprivation, starvation, and plant toxicities.
7. Incident clusters of plant toxicity (e.g., astragalus species), congenital or developmental abnormalities (e.g., parrot mouth, club foot) and unusual illness or death loss may need to be investigated on the range.
8. We need a better understanding of how subclinical conditions (e.g., nutritional deficiency, plant toxicity) impact immunocompetence and how animals cope with stress after removal
9. Procedures for blood banking need to be developed (also a recommendation of the WH&B Advisory Board) and implemented throughout the program.
10. There is no standard procedure for requesting, receiving, evaluating, and funding new research proposals. Unsolicited proposals should not drive the process. The BLM's needs should drive the process for research priority setting and calls for proposals.

Goals

11. Optimize preparation and health maintenance protocols to maintain, protect and improve the health and well being of wild horses and burros held in captivity.
12. Maintain safety for both the animals and persons involved during handling.
13. Be receptive to opportunities for improving handling procedures for wild horses and burros so they are better, and more efficient.
14. Better understand the health problems that occur in free ranging horses and how these problems may impact their health after removal.
15. Improve the adoptability of the wild horses and burros, and provide the public with animals that are healthier at the time of adoption, have lower rates of exposure to infectious disease and fewer health problems after adoption.

Strategies

The need for research to improve our understanding of the health needs of wild horses and burros on- and off-the-range is recognized in the context of the overall strategic plan for wild horse and burro research. This includes considering the impact of all research efforts (contraceptives, habitat evaluation, population modeling, etc.) on animal health and supporting research efforts aimed more specifically at animal health. Topics considered top priorities for health research include:

1. Consider additional research into the types of stress and injury that occur in wild horses and burros, their causes, and the best methods to reduce them during the transport. For example, identifying specific feed or trailer specifications that will minimize stress and injuries, and developing a better understanding of how electrolyte levels can be maintained during transport would be helpful.
2. Review the need for additional research on infectious upper respiratory diseases at holding facilities, identify sources of these infectious and ways to reduce the incidence of disease and impact on the program.
3. Develop a new more detailed monitoring system for mortalities and non-fatal illness and injury, as well as therapeutic and preventive treatments within facilities.
4. Research to better understand the incidence, causation, and prevention of diseases specific to wild burros.
5. Evaluate the effectiveness of vaccination strategies for strangles, equine influenza, herpes, and other diseases in stressed horses.
6. Review the need for investigating new gelding procedures, including the potential for injections of agents such as those that act against GnRH (Gonadotropin Releasing Hormone).
7. Develop and implement procedures for the safe, efficient banking of blood samples to facilitate animal health monitoring as a tool for improving animal health management (e.g. the emerging threat of West Nile virus infection).
8. Establish a standardized procedure for requesting, receiving, reviewing and funding health research proposals. The review procedure should include consideration of the scientific merits of the proposal as well as possible management implications for the BLM.
9. Use reviews of recent state-of-the-art research from the veterinary literature to address some of the Program's health research needs where possible. Make this information more readily available to specialists and thus, meet some information needs within the WH&B Program without the need for original research.

Other health problems and factors that are worthy of investigation in free roaming herds include problems or developmental abnormalities that may be related to nutritional deficiencies, plant toxicities or genetics, the unique attributes of wild horses that distinguish them from their domestic cousins (e.g., exceptional hoof quality), and monitoring wild horse and burro populations for infectious diseases that also impact the domestic horse population (e.g., vesicular stomatitis, equine infectious anemia).

Proposed outcomes

1. Minimize transportation stress and injuries.
2. Reduce or eliminate incidence of upper respiratory diseases at some facilities.
3. A new health monitoring system for wild horses and burros.
4. Improve health of wild burros and reduce mortality and incidence of serious illness that result from the special needs of burros.
5. Ensure the most up-to-date and efficient vaccination strategies are available to the program.
6. Identify the most humane, effective and efficient gelding procedures.
7. Increase availability of blood banking procedures for all blood samples taken.
8. A program for objective and fair calls for review, and funding of proposals.
9. Review all research proposals for their potential impact on animal health.

B. Fertility Control in Wild Horses - Strategies

The Challenge

No single fertility control agent or device is currently available that meets all of the stated needs of the BLM. Contraceptive agents or physical devices are currently available that meet many of the stated goals. Both Conventional PZP and Time-Release PZP technology meet more of BLM's needs than any other tested known agent. Although significant questions remain concerning population-level treatment and behavioral effects, both Conventional and Time-Release PZP are ready for field trial testing in wild populations under a research protocol. A longer-lasting agent is needed for larger herds to provide contraception with a single dose. New adjuvants need to be tested as potential replacements for FCA. Even as research continues and new management programs are instituted, the USGS-BRD and the BLM must remain alert to any additional, potentially promising new developments and new agents. The BLM and USGS-BRD should not immediately abandon any ongoing work with PZP until a significantly better agent, or better form of the existing PZP agent, is developed and tested.

Issues

1. Over the next five years, the BLM plans to lower populations of wild horses and burros to levels more compatible with a healthy range in order to (1) improve the condition and survival of the wild equids and other wildlife, and (2) protect soil and vegetation resources.
2. This goal is nearly impossible to obtain with removals and adoptions alone. The current existing adoption process cannot handle the 10,000 animals gathered annually during the last few years. There are limitations of both facilities and demand by the public to adopt animals.
3. Trade-offs between a fertility control agent's effective duration, cost, and risk of population over-management have not been quantitatively explored. The BLM needs to know the implications of variable durations of agent(s) on their program. Are two

agents clearly needed — one with short-term duration for small populations and one with longterm effects for larger populations?

4. To understand the effects of any stress due to handling for contraception.

Goals

1. Research should, as soon as possible, provide an effective fertility control tool to BLM for management use with wild horse mares that: (1) can be administered with a single injection; (2) is effective for multiple years; (3) is safe if administered to pregnant mares;
2. (4) is immediately reversible or passively wears off after which completely normal fertility and pregnancy in the mare can resume; (5) can be practically administered either remotely to habituated free-ranging wild animals or injected into animals held in facilities or restrained by equipment such as standard capture chutes; (6) will not enter the food chain; (7) can be tested under an INAD permit, or in the remote possibility of commercial interest, could eventually earn FDA approval; and (8) is greater than 90% effective in blocking pregnancy. At present, two forms of PZP – Conventional (multi-injection, one- year) and Time-Release (one-injection, two-year) –meet all of these criteria with one major exception; these forms do not meet the stated one-shot, three to five-year duration need of the BLM. Even the current Time-Release form needs some modification. Time- release PZP pellets made by either a heat-extrusion or a cold-evaporation process need to be evaluated. Beginning in 2002, field-trial tests were initiated with the currently available PZP agents. These will continue until a longer-lasting agent can be developed or becomes available. The purpose of these field trials is to provide the necessary information and assurances on population-level and behavior effects so that broad scale management of herds by PZP may proceed.
3. Develop a longer-lasting (three to five, or more, years) agent, or further modify and extend Time-Release PZP to achieve the one-shot, three to five-year need of the BLM. A longer-lasting agent must be developed and tested by the USGS-BRD and the BLM. Even a one-year extension of the current Time-Release form to a three-year duration would make a huge savings of efforts and costs for managers of larger herds. Given the fact that many gathers (and thus, the injections) would occur during the summer or early fall, the current two-year time-release form only provides contraception for one full breeding season. Even two breeding seasons of fertility control may be too brief for most herd managers since most gathers are on a three- or four-year cycle. In light of this, BLM initiated a test of a 3-4 year PZP vaccine with captive wild mares in March 2005.
4. Test alternative adjuvants to FCA that do not result in any false positive TB tests and are less objectionable. Test MFA, QS-21, and other adjuvants that would be less objectionable than FCA. A 10-month trail did occur with MFA in captive wild mares and was completed in October 2004. MFA is now being considered as an acceptable substitute to FCA. BLM will consider the replacement of FCA with MFA in all subsequent captive and field trials as the adjuvant to be used with PZP.

Strategies

1. Conduct preliminary modeling for direction and guidance on the numbers, age classes, and durations of treatments. — Model the tradeoffs, cost-benefits, and risks of the various durations of an agent. Based on this analysis, the BLM should then select the optimum duration(s) for agents that they desire to use. Modeling should also guide the optimum mix of strategies (all contraception, all removal, or best mix of both), scenarios of marking, monitoring, handling, and mare selection (all young, random) to produce the greatest benefits to handlings and animals removed, to meet BLM's stated objectives for population viability. This preliminary modeling began in 2002, and is being done by researchers at Colorado State University. In addition, an economic analysis evaluating different management strategies with fertility control was modeled by John Bartholow, a BRD-USGS researcher, and completed in 2004.
2. Immediately conduct field tests with both Conventional PZP and Time-Released PZP under a research protocol until a longer-lasting agent is developed (Figure 2). Conduct research on the effects of PZP contraception on wild horse population growth rates, seasonality of foaling, any health complications, and any effects on behavior, and harem dynamics so that the information and assurances are available prior to proceeding with any broad-scale management application of PZP. Conduct those studies on wild horses in captivity for those topics where free-ranging wild horses cannot be observed with enough regularity (e.g., studies of effects on annual estrus cycling, studies of any complications). Population and individual-based studies, using conventional and time- release PZP, were initiated in 2002 and 2003 under the guidance of the Fertility Control Field Trial Plan.
3. Organize a competitive call for proposals to develop a longer-lasting agent.
4. Preliminary laboratory work with mixtures and dosages should occur by the developers, proponents and contractors.
5. Screen the safety of any new potential longer-lasting contraceptive agents first in captive settings before any testing in the field. Included in this listing are any new agents, adjuvants, and mixtures. Tests of FMA and a newly-developed 3-4 year PZP vaccine have been initiated with captive wild mare trials in 2004 and 2005, respectively.
6. Immediately start tests on the efficacy and duration of other adjuvants that are less objectionable and have a higher probability of FDA approval compared with FCA (e.g., Modified Freund's, QS-21)
7. Consider the potential for FDA-approvability in the selection of agents. If a longer-lasting (3-4 year) agent is developed, field trials on that agent should be immediately initiated.

The BLM has selected a two-pronged plan of attack to simultaneously: (1) aggressively pursue a final form of new agents or modification of currently available PZP that meets the longer-term duration needs, while also (2) aggressively field testing current PZP until any significantly better agent becomes available. Work on development and screening of new agents (Phase II) and field testing the best current agents (Phase III) should occur concurrently (Figure 2).

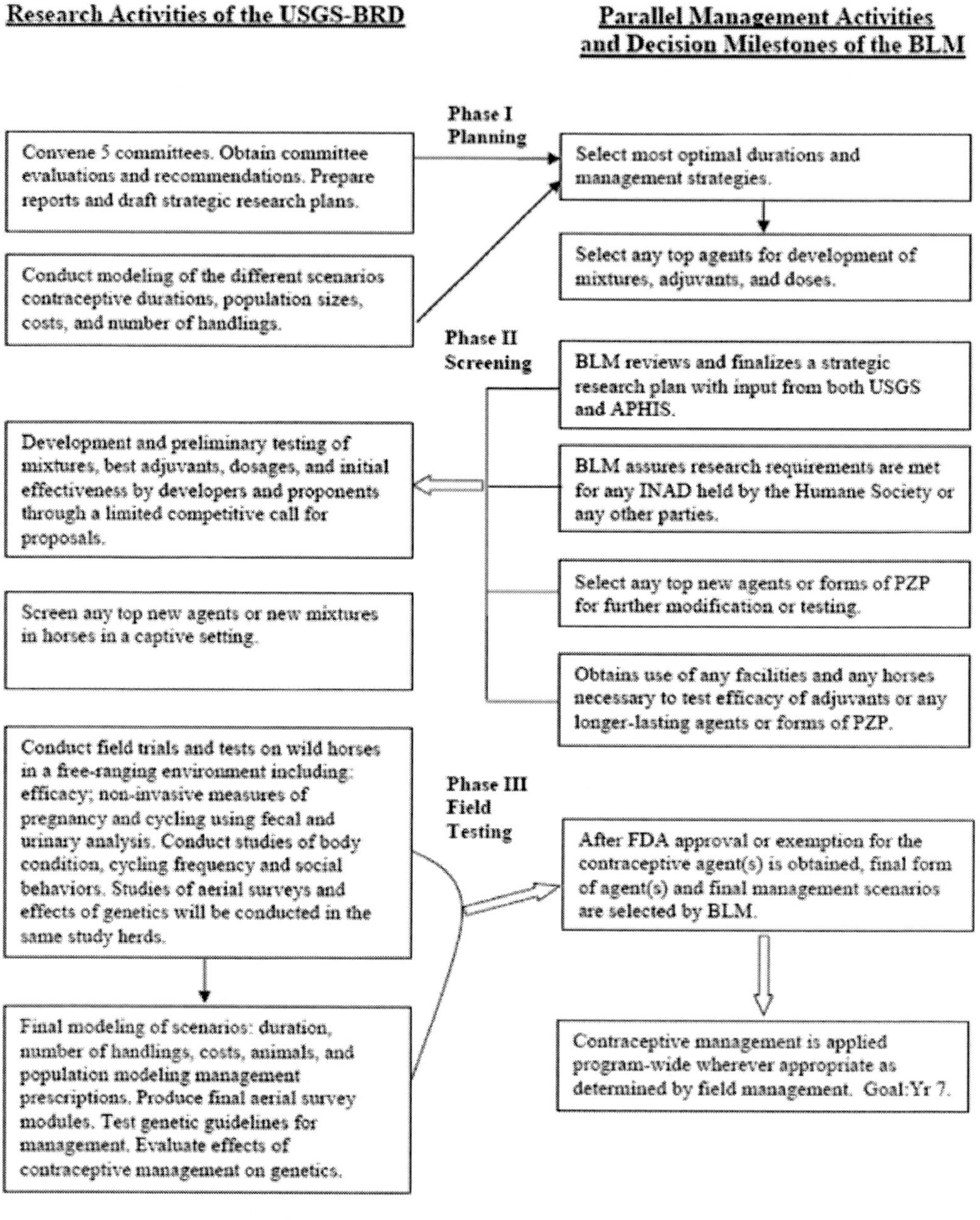

Figure 2. Project overview. Major research steps and corresponding decision milestones for the BLM are detailed.

The BLM and USGS-BRD initiated field tests with current PZP forms in 2002 and 2003, under the guidance of the Fertility Control Field Trial Plan. The respective agency research coordinators traveled to prospective study sites in 2002 and 2003 and visited with the herd area managers. Study sites were selected and field research initiated to obtain pre-treatment data wherever possible. Field studies started in 2002 and 2003, to run through 2007 and 2008, yielding one to two years of pretreatment data and four years of post-treatment data. Where pretreatment data collection was not possible, the controls to treated mares will be untreated mares in the same and in other harems. Research with aerial survey techniques and any field studies related to genetics should be conducted on these same study herds in order to maximize efficiency and save dollars. The field-trial research with PZP will also include the agent's effects on fertility control and body condition, studies on age-specific survival rates, harem organization and cohesion, duration of the breeding season, social behavior of mares, harem-tending behavior, and dominance relations among stallions.

Proposed Outcomes

1. A suitable, longer-lasting, and safe fertility control agent will be developed for consideration in broad-scale management application by the BLM. Fewer total animals will be rounded up and adopted. When lower maintenance population levels are reached, there will be longer intervals between gathers, and the gathers will be less expensive and smaller in scale. Fewer unadoptable animals will be held in long-term holding facilities at government expense.
2. Substantial savings in federal dollars will be accrued to the government. Presently, a large proportion of the entire budget of the WH&B Program is tied up in the enormous "Adoption Pipeline" – the process of gathering animals, holding them, transporting them, and then adopting them.
3. More stable populations will be achieved with contraceptive management than with the widely fluctuating current "increase-gather-increase" scenario. More genetic heterozygosity will be maintained and habitat will be protected under a more stable population scenario.
4. Fewer animals will be subjected to the stresses of capture and increased diseases and pathogen transmissions that go along with confinement of large numbers of animals.

C. Population Estimation and Modeling of Population Management Prescriptions – Strategies

The Challenge

The stated goal for the BLM's Wild Horse and Burro Program is to conduct a population census[9] or estimate on every herd area every four years, or more frequently if necessary. A few of the smallest, most accessible, and most visible of the wild horse herds may be completely censused using ground surveys, with identification based on photographs or unique natural markings of animals. Because wild burros possess fewer unique markings, identification systems do not appear to be possible for wild burros. The size of all burro herds and most wild horse herds are estimated from an aircraft, typically a helicopter. Wild horses in flat, treeless terrain are easily counted and it is estimated that ~ 85% of animals are seen. But, wild horses and burros may be missed in more rugged terrain and tree cover. Wild horse and burro specialists need standardized, tested, cost effective, defensible, yet easy-to-use aerial population estimation techniques for wild burro herds found in rugged and forested

[9] A census is typically defined as a total count. Total counts are rarely possible, however, since animals are found in tree cover and dispersed over large areas of rough terrain. Thus, a population estimate with a confidence interval on that estimate is a more reasonable goal for managers.

areas. The best technique(s) presented to the BLM wild horse managers in a format such as a computer diskette or CD-rom with easy-to-follow directions and a user manual. Some areas to be surveyed are so vast BLM managers may decide to count only portions of a herd area to save time and dollars. Stratified, random subsampling procedures may be used to obtain a valid estimate of the population size for these larger herds.

Issues

1. Aerial and ground population survey requirements for the BLM are daunting. The BLM is responsible for the management of over 200 wild horse and burro populations located across vast expanses of public lands.
2. In order to plan any management removals, the BLM requires population estimates, sex and age classification data, and population management modeling (Jenkins 1996) every three years on every herd area.
3. Sightability for wild burros is lower due to their small size, often cryptic colors, small groups (sizes are often single individuals), and stoic behavior. This results in many animals standing still as aircraft pass the rugged, brushy or riparian-forested habitats that they reside in. Sightability for burros from a helicopter may only be 40-60% of the animals present.
4. Wild horses are visible in flat, treeless terrain, but where patches of conifer cover are present, the estimated sightability of wild horses may be as low as one-half the animals present. Even in the more open terrain in Nevada, earlier scientific estimates of the percent of all wild horses observed are still only 66 to 85%.

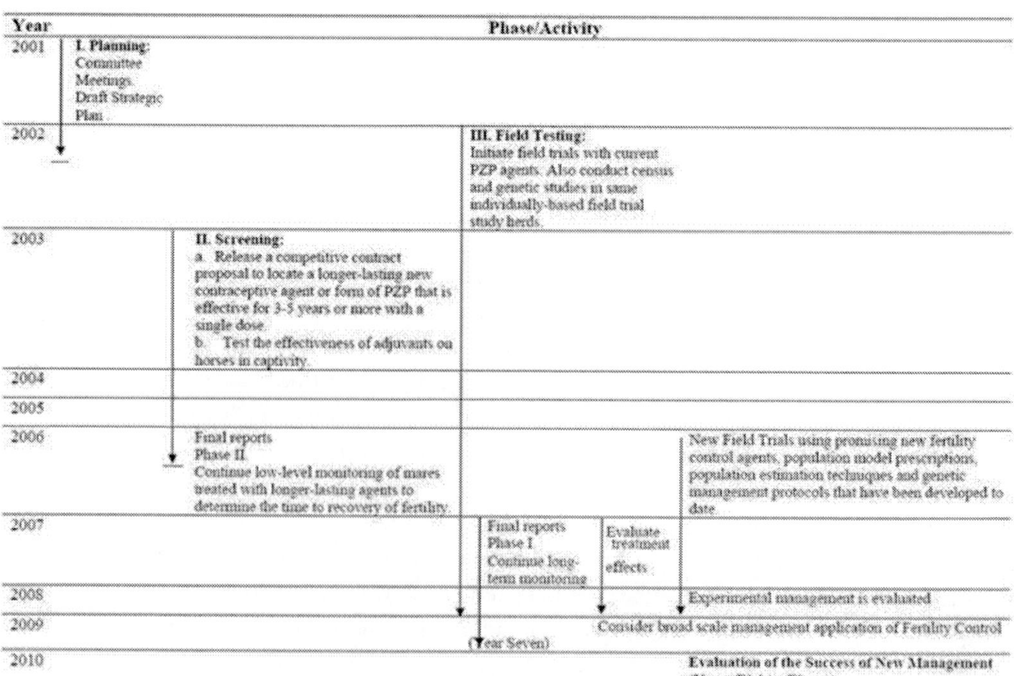

Figure 3. Phases I-II of the Strategic Research Plan for Wild Horses and Burros.

5. Cost, personnel, and fatigue factors may make aerial surveys of large areas prohibitive. Representative samples of these vast areas can be surveyed using stratified, random sampling, thus reducing the total area that needs to be surveyed. The BLM has directed the protocol include the option for subsampling, using these procedures.

Goals

1. To provide aerial and ground population estimation techniques for wild horses and burros that are valid, tested, defensible, cost-effective, easy to use, and that allow managers to survey representative reproductive subsamples of vast areas to obtain valid population estimates. Testing on several recommended techniques began in 2004.
2. To explore high resolution Department of Defense imagery as a possible substitute for helicopter surveys, if the technology was made available to the BLM.
3. To improve, update, and apply current population and genetic models to guide removal, genetic and contraceptive management needs.

Strategies

1. Develop aerial technique(s) and product(s) from amongst four candidates (Idaho Sightability Model, Simultaneous Double-Count, Distance Sampling, and Noninvasive Mark-Resight) for wild horses. Many wild horse herd management areas are managed for small populations that may be more amenable to mark-resight techniques using individually identifiable horses, and representative ground or aerial surveys of the area. Focus on development of those least invasive techniques that: (a) require no collaring or capture of wild horses, (b) only one aerial or ground survey per population estimate to minimize overflights and disturbances of the animals, and (c) allow subsampling.
2. Explore the availability of Department of Defense (DOD) satellite imagery for identification of horses as a potential substitute to helicopter surveys in some areas. However, the war on terrorism may make this imagery less available for the next few years.
3. Continue to monitor the progress of the Arizona Wild Burro Interagency Aerial Survey Working Group to develop a technique for wild burros. Currently the group is testing the Simultaneous Double-Count technique, with excellent initial success. Respond if the group asks for additional work or assistance. The need for wild burro aerial survey techniques will be addressed following the wild horse work.
4. Three population models have been developed for management application, all of which received high marks for performance by this and earlier panels. These models should continue to be updated and improved. In particular, the Jenkins population-removal model (Jenkins 1996) that is used by wild horse and burro specialists, is currently being revised for use with WINDOWS, and will include density dependence —both are significant improvements. The Gross model (Gross 2000) is excellent for modeling genetics and management scenarios. The Hobbs Model

(Hobbs et al. 2000) is excellent for modeling contraceptive scenarios — but both models need updating with new research information.

Proposed outcomes

1. Improved methods for population estimation will result in aerial surveys in herd management areas that are less expensive, more efficient and provide accurate estimates with confidence intervals on the estimates.
2. The techniques will be accessible by all wild horse and burro specialists.
3. The Jenkin's population model that the BLM currently uses will be improved.
4. Population models will be applied to more management questions concerning genetics and fertility control treatment plans.
5. DOD satellite imagery could substitute for helicopter surveys. Considerable cost savings could occur as could reduction in safety hazards to BLM personnel, if this imagery could be made available.

D. Genetic Conservation Strategies

The Challenge

BLM is mandated to manage for self-sustaining populations of wild horses and burros. Inbreeding appears rare in wild horses and burros. Genetic problems due to inbreeding depression, however, may be encountered in a few small, isolated populations of wild horses or wild burros. The BLM needs to guard against potential inbreeding problems by conducting surveys, as needed, of the genetics of wild horses and monitoring key indicators of heterozygosity, inbreeding coefficients, and genetic effective population sizes so that management intervention may be proactive.

Some potentially unique groups and phenotypes of wild horses occur on the BLM lands. The public recognizes these unique groups, particularly a few herds with Spanish colonial heritage. Existing genetic evidence supports the presence of early colonial Spanish horse alleles in some herds. This suggests these herds are more similar to early Spanish founder stock. However, wild horses possess no unique alleles that are not already found in domestic horses. Nothing in the 1971 Act or in policy, however, directs the BLM to provide special management for some groups of horses. The genetic and heritable components of any possibly unique traits, or unique groups of wild horses, should be tested during a comprehensive analysis of common ancestries amongst the herds. Similar or closely related herds of horses should be identified for any genetic augmentation of wild horse herds.

Issues

1. Population goals for some wild horse and burro herds may be too low to meet conventional standards for minimum genetically viable sizes (K. Schoenecker and F. Singer 1999. USGS-BRD report, Ft. Collins, CO). This is not a matter of immediate concern since many of these herds may have gene flow to other herds, thus forming a metapopulation. Even very limited gene flow (e.g., one to two breeding animals

every generation) between subpopulations will guard against inbreeding. Wild horse herds were larger in the recent past, peaking in numbers about 1980. Thus, genetic concerns are a fairly recent issue. Only approximately five herds have produced animals with physical defects. In only one of these cases was there sufficient background information on the herds' size and genetics to implicate inbreeding as the likely cause of the problems. Inbreeding may have caused the other problems, but they may also have been due to inherited defects.

2. Managers lack data on which herds are genetically isolated. Those herds need to be identified that possess low values of heterozygosity, and where severe bottlenecks, or high levels of inbreeding occur.
3. Unique phenotypic, historical and wild types of horses may exist; however, the genetic basis for any potential uniqueness has never been quantified.
4. Wild horses may be more vulnerable than many mammals to inbreeding depression at low population levels due to: (1) a harem breeding structure that limits breeding males mostly to harem holding stallions, and (2) a dominance hierarchy that usually delays harem holding and breeding in males until six to seven years of age or older.

Goals

1. Manage against inbreeding depression. Maintain healthy genetic fitness and viability of wild horses as a wildland species so they may survive and persist in unpredictable and often rigorous environments. Manage to minimize the need for augmentations, if possible.
2. Set minimum goals for genetic viability of the populations in terms of: (a) minimum levels of total heterozygosity (both allozyme and DNA), (b) minimum genetic effective population sizes, (c) maximum amounts of inbreeding, and (d) a maximum loss of alleles that will be allowed.
3. Identify any small[10] and truly isolated herds of wild horses through both genetic analysis and also reviews of the movements, herd histories, and sizes of the herds in question. Small and isolated herds are a cause of concern. Identify any metapopulations or collections of connected subpopulations where small size of some subpopulations will not be a concern.
4. Document any relatedness and any uniqueness of all the herds. Quantify relatedness amongst herds to guide any introductions of new animals. Establish prevalence and habitability of any special traits or phenotypes.
5. Establish management protocols for genetically "rescuing" small, isolated populations of wild horses through introductions of new individuals from genetically related or similar herds.
6. Conduct research on what is the minimum size for viable populations.

---

[10] Small populations are defined as $N_e < 50$.

Strategies

1. Conduct a comprehensive survey of the genetics of all the wild horse populations. Determine which small wild horse herds, if any, are truly genetically isolated. Although
2. previous genetic blood draws were done on several herds, systematic surveys were initiated in 2003 whereby all herds subjected to population control gathers were also subjected to blood draws for genetic evaluation purposes. Reports are generated for each herd identifying founder stock, level of diversity and management recommendations for avoiding inbreeding problems.
3. Determine the effects of contraception on rates of loss of herd heterozygosity.
4. Document the inbreeding genetic contribution, if any, to the deformities observed in a few specific herds. Develop guidelines for avoiding these situations in the future, and for avoiding deformities that have a genetic component.
5. Determine, through field studies, the threshold levels of heterozygosity and inbreeding below which reduced fitness and inbreeding depression may occur in wild horses, so that the BLM can avoid these situations.
6. Calculate genetically effective population sizes (Ne) in at least five populations of wild horses (two are already completed) so that the BLM will know the average and range of Ne to census N ratios to establish generality for minimum population sizes for viability.
7. Conduct a meta-analysis of the genetic information for all the herds. Conduct quantitative statistical analyses to identify the commonalities and relatedness of all the groups of wild horses. These commonalities may guide which herds are selected for the introduction of new animals. Additionally, the thresholds for management intervention need to be quantitatively assessed, defensible and published. Currently, the genetics contract only allows for sampling and analysis of individual herds
8. Search for any patterns in fitness lines, matrilines, patrilines, etc., that would alter the estimation of Ne/N. How managers select horses for removals vs. those left on the range to breed may alter genetics of the herd. Conduct an analysis of parentage, using DNA markers, in three individually-based study herds to accomplish this.
9. Identify meta-populations within each given area, region, or state.

Proposed outcomes

10. Minimum guidelines and protocols will be established for management intervention to maintain fitness, avoid inbreeding depression, and prevent genetic defects due to inbreeding in wild horse herds. Key genetic parameters will be monitored.
11. There will be documentation of the genetic basis for the relatedness of groups to guide the introduction of new animals for genetic purposes to identify any unique traits or any unique groups of wild horses.

E. Habitat Assessment and Setting Population Goals

Deferred pending a more in-depth analysis of BLM's needs related to habitat assessment, monitoring and evaluation.

IV. GLOSSARY

(a) *Adjuvants* - A component of the vaccine that enhances the main product, or antigen, and consequently increases antibody formation. Adjuvants can also delay the release of the vaccine as in the case of SpayVac. The vaccine works much more efficiently with the adjuvant. These are general immunostimulants that cause the body to make greater concentrations of antibodies against the vaccine.

(b) *Appropriate Management Level (AML)* – The Wild and Free-Roaming Horse and Burro Act of 1971, the Taylor Grazing Act of 1934 and the Federal Land Policy and Management Act of 1976 all direct the BLM to set appropriate numbers of grazing animals, including wild horse and burros, and to manage for those goals, or AMLs. AML goals are determined through the BLM's planning process, and are later evaluated with monitoring at levels that will allow for healthy, self-sustaining populations of equids, a Thriving Natural Ecological Balance (TNEB), and are compatible with other uses.

(c) *Chemosterilants* - This term broadly refers to any chemical agent that causes temporary or permanent sterility.

(d) *Contraception* - A chemical or agent that prevents pregnancy in any manner, typically by blocking either ovulation or fertilization. Contraception is usually temporary. Fertility usually returns passively after treatment has been terminated.

(e) *Fertility Control* - A collective term that refers to all methods of inhibiting reproduction.

(f) *Immunocontraceptives* - Contraceptive agents stimulate the body's immune response in the host animal against hormones or proteins essential for reproduction, and in doing so blocks pregnancy or some other essential component of reproductive function. Immunocontraception may include vaccines directed at either reproductive hormones, at sperm, or at the ovum. Most wildlife applications include vaccines that are directed at blocking fertilization in the female by stimulating production of antibodies against the zona pellucida (ZP) of the ovum.

(g) *Investigational New Animal Drug Exemption (INAD)* - A research permit issued by the U.S. Food and Drug Administration (FDA) for research into new or untested compounds including contraceptive agents. This permit is required for any new drug research with animals.

(h) *Passive Return, Reversible and Permanent Fertility Control Agents* - Reversible fertility control agents as those agents whose antifertility effects can be immediately terminated with a reversing agent and normal reproductive function resumed. Immediate reversibility might include a reversing agent that is administered, or a physical device such as an IUD that can be removed and normal reproductive function returns. A reversible agent provides a number of advantages in that the number of animals returned to reproductive potential is immediately known. PZP is an example of the agents that, in the absence of an annual booster, normal reproductive function is passively resumed, as the effects of the agent wear off. The term permanent when applied to fertility control agents is also operationally vague. Except for the removal of the reproductive organs which permanently impair reproduction, the agents often referred to as permanent, such as GnRH-PAP, may be long-term but may not prove to be truly permanent.

(i) *Physical Methods of Fertility Control: Intrauterine Devices (IUDs)* - Physical rings or other devices that are placed in the uterus and prevent pregnancy. This approach has been used to successfully block pregnancy in humans and in a large number of animal species, including six domestic horse mares (Daels and Hughes 1995). All of these mares produced foals following the removal of a O-ring shaped IUD. In humans, IUD's were demonstrated to be effective for as long as 12 years, at which time their effects were still reversible.

(j) *Sterilization* - An inability to reproduce. Usually refers to a more or less permanent infertility, such as would occur by removal of male (castration, or the gelding process) or female reproductive organs (neutering, or spaying process). Several agents listed in this report mimic sterilization in the short term, or in some cases longer term, e.g., seven years, but none are felt to cause permanent sterility – some recovery of normal reproductive function is likely.

(k) *Strangles* – Strangles is a highly contagious, infectious respiratory disease of equids caused by the bacterium *Streptococcus equi* subspecies *equi.* Characterized by inflammation of the upper respiratory tract infection often results in abscessation of intermandibular, parotid or pharyngeal lymph nodes. Immunologically naïve horses are at greater risk for infection and outbreaks often occur in large populations that have frequent additions. Although mortality is rare, morbidity can approach 100%. Immunization against *S. equi equi* does not always prevent infection, though it may prevent severe manifestations of disease in infected animals.

(l) *Thriving Natural Ecological Balance (TNEB)* - The Wild and Free-Roaming Horse and Burro Act of 1971 requires that wild horses, burros and wildlife, be in good health and reproducing at a rate that sustains the population; the key vegetation is able to maintain its composition; the soil is being protected; and a sufficient amount of high quality water is available to all animals.

ACKNOWLEDGMENTS

The efforts of Linda Coates-Markle (BLM), Albert Kane (APHIS) in reviewing all sections of this plan are appreciated. Dr. Kane drafted the health sections and Francis Singer (BRD) the remaining sections. The help of Tom Pogacnik, John Fend, Bud Cribley (BLM), Lisa Hatcher (APHIS), and Sue Haseltine, Denny Fenn, Bernie Shanks, Doug Posson, Juliette Wilson, Greg Smith, and Zachary Bowen (USGS-BRD) was invaluable to the planning process. Matt Tobler, Laurie Richards, and Jennifer Sunderland assisted with the many redrafts of this plan. The members of the five advisory committees greatly assisted this planning effort: The Health Committee: Katie Blunk, APHIS; Linda Coates-Markle, BLM;

Ann Davidson, Louisiana State University; Ted Friend, Texas A&M University; Lisa Hatcher, APHIS; Albert Kane, APHIS; Sharon Kipping, BLM; Robin Lohnes, WHB Advisory Board; Rebecca McConnico, Louisiana State University; Mo Salman, CSU; Francis Singer, USGS; Mark Stewart, APHIS; Joe Stratton, BLM; Josie Traub-Dargatz, CSU; Bea Wade, BLM. The Genetics Committee: Linda Coates- Markle, BLM; Gus Cothran, University of Kentucky, Philip Hedrick, Arizona State University; Leonard Nunney, University of California; Sara Oyler, USGS; Bruce Pugesek, USGS; Rob Ramey III, Denver Museum Of Nature and Science; Dan Rubenstein, Princeton University; Philip Sponenberg, Virginia-Maryland College of Veterinary Medicine. The Contraceptives Committee: Lowell Miller, APHIS Co-chair; Dan Baker, Colorado Division of Wildlife; Linda Coates-Markle, BLM; Gary Killian, Pennsylvania State University; Patrick McCue, Colorado State University; Glen Sargeant, USGS; Mike Whitacre, University of North Carolina. The Population Committee: Linda Coates-Markle, BLM; E.O, Garton, University of Idaho; Kelly Grissom, BLM; Lisa Hatcher, APHIS; John Hervert, Arizona Game and Fish; Tom Hobbs, CSU; Steve Jenkins, University of Nevada; Bruce Lubow, APHIS; Glen Sargeant, USGS.

REFERENCES

Bennett, D. & Hoffman, R. S. (1999). Equus caballus. *Mammalian Species*, *628*, 1-14.

Bennett, D. (1998). *Conquerors, the roots of New World horsemanship*. Amigo Publications, Solvang, California., 422.

Bowling, A. T. & Touchberry, R. W. (1990). Parentage of Great Basin feral horses. *Journal of Wildlife Management, 54*, 424-429.

Brown, R. G., Bowen, W. D., Eddington, J. D., Kimmins, W. C., Mezei, M., Parsons, J. L. & Pohajdak, B. (1997a). Temporal trends in antibody production in grey harp and hooded seals to a single administration immunocontraceptive vaccine. *Journal of Reproductive Immunology, 35*, 53-64.

Brown, R. G., Bowen, W. D., Eddington, J. D., Kimmins, W. C., Mezei, M., Parsons, J. L. & Pohajdak, B. (1997b). Evidence for a long-lasting single administrative contraceptive agent in wild grey seals. *Journal of Reproductive Immunology*, *35*, 43-5 1.

Burnham, K. P., Pugesek, B., Sargeant, G. A., Singer, F. J., Underwood, H. B., Shanks (compiler), B. (2000). *USGS Review of Wild Horse Population Model,* USGS-BRD, Washington D.C. 2.

Coughenour, M. B. (2000). *Ecosystem modeling of the Pryor Mountain Wild Horse Range.* Final Report to U.S. Geological Survey and Bureau of Land Management, Colorado State University, Fort Collins.

Curtis, P. D. & Warren, R. F. (1999). *A workshop on the status and future of wildlife fertility control.* The Wildlife Society, Washington, D.C.

Floorman, P. M. & Wasserman, H. M. 1985. O-linked oligosaccharides of mouse egg ZP3 account for its sperm receptor activity. Cell, *41*, 313-324.

Garrott, R. A. & Siniff, D. B. (1992). Limitations of male-oriented contraception for controlling feral horse population. *Journal of Wildlife Management, 56*, 456-464.

Gross, J. E. (2000). A dynamic simulation model for evaluating effects of removal and contraception on genetic variation and demography. *Biological Conservation, 96*, 319-330.

Gross, J. P., McCue, L., Miller, O., Ryder, & Underwood, B. H. (B. Shanks, compiler). 1999. Research evaluation: Contract for immunization/contraception and fertility in wild horses. USGS-BRD Report to BLM.

Hobbs, N. T., Bowden, D. C. & Baker, D. L. (2000). Effects of fertility control on populations of ungulates: General, stage-structured models. *Journal of Wildlife Management, 64*, 473491.

Holzgang, O. (1997). Herbivore – carrying capacity of grasslands in the Swiss National Park. Thesis. *Swiss Federal Institute of Technology*, Zurich.

Jenkins, S. H. (1996). *Wild horse population model, Version 3.2.* Users Guide. Department of Biology, University of Nevada, Reno.

Johnson, B. K., et al. (2001). Application of multi-species stocking allocation models to study herbivory in the interior West USA. Paper presented at *International Conference on Forest Dynamics and Ungulate Herbivory,* Davos, Switzerland, October 3-6, 2001.

Kirkpatrick, J. (1995). Management of wild horses by fertility control: the Assateague experience. *NPS Scientific Monograph No. 26*, Denver Colorado.

Kirkpatrick, J. F. & Turner, A. (2002). (in press). Reversibility of action and safety during pregnancy in wild horses immunized against porcine zona pellacida. *Reproduction.*

Kirkpatrick, J. F. & Turner, J. W. (1982). Reproductive biology and chemical fertility control in feral horses. *Final Report to BLM on Contract.*

Kirkpatrick, J. F. & Turner, J. W. Jr. (1991). Changes in herd stallions among feral horse bands and the absence of forced copulation and induced abortion. *Behavioral Ecology and Sociobiology, 29(3)*, 217-219.

Kirkpatrick, J. F., Liu, I. K. M., Turner, J. W. Jr., & Keiper, R. (1992). Long term effects of porcina zonae pellucida immuno contraception on ovarian function of feral horses. *Journal of Reproduction and Fertility, 94*, 437-444.

Kirkpatrick, J. F., Shideler, S. E. & Turner, J. W. Jr. (1990). Pregnancy determinations in uncaptured feral horses based on free steroids in feces and steroidal metabolites in urine-soaked snow. *Canadian Journal of Zoology, 68*, 2576-2579.

Kissell, R., Jr., Irby, L. R. & Mackie, R. J. (1996). Competitive interactions among bighorn sheep, feral horses, and mule deer in Bighorn Canyon National Recreation Area and Pryor Mountain Wild Horse Range. *Completion report of the U.S. National Park Service.* Montana State University.

Kreeger, T. editor. (1997). Contraception in Wildlife Management. USDA, Animal and Plant Health Investigation Service. *Technological Bulletin*, 1853.

Little, D. K., Grissom, K., Oyler, R., Barnes, C., Elefritz, S., Christensen, J., Acheson, G., Thomas, L., Herrert, J., Lee, R., Fuller, A., Pierce, M. & Kohnke, S. (1999). Report of the Arizona Burro Census Team. BLM, Phoenix, *Arizona.*

Liu, I. K. M., Bernoco, M. & Feldman, M. (1989). Contraception in mares heteroimmunized with pig zonae pellucidae. *Journal of Reproduction and Fertility., 85*, 19-29.

National Research Council. (1982). Wild and free-roaming horses and burros. Final Report. National Academy Press, Washington, D.C.

National Research Council. (1991). *Wild horse populations: field studies in genetics and fertility.* National Academy Press, Washington, D.C. 42.

Powell, D. M. (1999). Preliminary evaluation of porcine zona pellucida (PZP), immunocontraception for behavioral effects in feral horses (Equus caballus). *Journal of Animal Welfare Science, 2*, 321-335.

Schoenecker, K. A. & Singer, F. J. (1999). A survey of BLM wild horse managers to assess research needs. Report to the BLM. USGS, *Midcontinent Ecological Science Center., 8.*

Smith, G., Bacomb, C., Estielman, K., Aicher, D., Harrison, K. & Little, D. (1996). Report of the review team on forage allocation for wild horses and livestock. USDI, *Bureau of Land Management*, Reno, NV. 37.

Turner, A. & Kirkpatrick, J. F. (2002). (in press). Effects of immunocontraception on population, longevity and condition in wild mares (Equus caballus). *Reproduction.*

Turner, C. L., Seastedt, T. R. & Dyer, M. I. (1993). Maximization of above ground grassland production: The role of defoliation frequency, intensity, and history. *Ecological Applications, 3*, 175-186.

Turner, J. W. (1999). Project Director. 2000, 2001, 2002. Annual Progress Reports on USDI-BLM Assistance Agreement. *Nevada Wild Horse Fertility Control Project.* Medical College of Ohio.

United States Department of the Interior, Bureau of Land Management. 2000. *Summary Recommendations – BLM Wild Horse and Burro Population Viability Forum,* April 21, 1999. BLM Resource Note No. 35, BLM, National Science and Technology Center, Washington, D.C. 4.

USDI, BLM. (1997). *Summary Recommendations – BLM Wild Horse and Burro Population Viability Forum*, BLM National Resource Note No. 35, National Science and Technology Center, BLM, Washington, D.C. 4.

Wagner, F. H. (1983). Status of wild horse and burro management on public rangelands. *North American Wildlife and Natural Resources Conference*, *48*, 116-133.

INDEX

A

B

C

D

E

F

G

H

I

J

K

L

M

N

O

P

Q

R

S

T

U

V

W

X

Y